SPARK, STREAMER, PROPORTIONAL AND DRIFT CHAMBERS

Spark, Streamer, Proportional and Drift Chambers

Peter Rice-Evans
Lecturer in Physics
Bedford College, University of London

THE RICHELIEU PRESS
LONDON

Published by
THE RICHELIEU PRESS LIMITED
30 Saint Mark's Crescent, London N.W.1

ISBN 0 903840 00 6
L.C. No. 73-90334

© 1974 The Richelieu Press Limited

All rights reserved. No part of this publication may be reproduced, stored in a retrieval system, or transmitted, in any form or by any means, electronic, mechanical, photocopying, recording, or otherwise without the prior permission of the publisher.

This book has been set in Times 327–11 point
and printed in Great Britain by
Fletcher & Son Ltd, Norwich

To my father
ELWYN
who taught me to think

Preface

Experimental physics is a privileged and yet a tantalizing pursuit. Like the sculptor chiselling a block of marble to discover his truth, the physicist discloses his by composing an experiment. With a conjunction of wires, gases, particles, currents and materials, he seeks to expose the natural laws of the Universe.

There is the end—the knowledge itself—and there is the means of obtaining this end. Views differ on which is the more important but for myself the very act of experimentation is the highlight of existence. It is, however, an experience that is in danger of disappearing; physicists no longer build their own accelerators, they do not extract their own beams, nor do they construct their own bubble chambers. But—fortunately I would say—one still cannot buy commercial spark chambers, nor can one expect them to be produced as standard units on request at a synchrotron laboratory. The prospect therefore always exists that, by ingenious design, the experimental art may be advanced by the practitioner to a stage where new phenomena are resolved. And so it remains possible for men to feel the elation and despair that accompanies the activity called counter physics. This monograph is written to serve such men.

The book is intended primarily for beginners—for, say, postgraduates commencing their designs. Although there is no substitute for recourse to the learned journals when it comes to the detailed planning of experiments, there is certainly a need for a single volume in which much of the basic material is collected and displayed. The newcomer to the field may then cast his eye over the pages to ascertain some of the difficulties of any undertaking he proposes.

In a book of reasonable size, a lot of material has to be omitted. Theoretical treatments have been restricted to the basic principles, but on occasion—where important—considerable practical detail has been included, especially if it happened to lie in some arcane preprint or internal laboratory report. Inevitably the choice of material has been influenced by my own prejudices, by what I have learnt with our streamer chambers at Bedford College, and by my experiences and discussions at CERN, Geneva, where the bulk of this monograph has been written.

In the course of writing, the subject has advanced remarkably. Spark chambers are still the mainstay of particle experiments but many other varieties of chamber have come forward. The streamer chamber has received general recognition as a powerful tool that many would say outrivals the bubble chamber. The multiwire proportional chamber has been introduced and large-scale systems are being brought into commission. In the last year or so the virtues of drift chambers have become apparent and they are sure to become important in the future. It is the author's hope that this book lays a satisfactory foundation to the understanding of this instrumentation.

Not surprisingly, the main applications of detection chambers have been in the realm of particle physics. Nevertheless Chapter 9 illustrates the wide variety of uses to which the chambers have been put. These range from X-ray and gamma-ray astronomy, to nuclear medicine and biology, to radioactive studies, and even to archaeology. And, of course, the devices open up important new areas for research into electrical discharges.

In writing this review monograph I have tried to select the most important elements of the subject. Inevitably, therefore, I am indebted to those whose work I have described. Some major contributions loom large and my indebtedness is consequently in proportion. Many friends, colleagues and professional acquaintances have helped me directly; and although it is hardly practical to itemize their assistance I gratefully acknowledge that I have benefited from collaboration, conversations, correspondence and other contacts, with the following: Drs. M. G. Albrow, L. W. Alvarez, P. Astbury, D. P. Barber, E. H. Bellamy, B. Bošnjaković, J. R. Brookes, R. C. A. Brown, D. Butt, G. Charpak, J. W. Cronin, S. Derenzo, Z. Dimčovski, G. Dowty, W. Evans, C. Grey Morgan, F. F. Heymann, T. W. Jones, Maria Kienzle, W. Kienzle, R. Klanner, F. Krienen, A. Ladage, R. F. Mozley, C. Nef, T. A. Nunamaker, G. C. Laverriére, V. I. Little, V. Perez-Mendez, F. Piuz, H. Raether, F. Rohrbach, V. N. Roinishvili, J. Saudinos, F. Schneider, P. Schübelin, H. Schmied, M. Steuer, L. Sulak, D. Sutherland, P. T. Trent, H.

Verweij, F. Villa, A. H. Walenta, D. Websdale, A. Weitsch, G. Erskine and W. Risk.

For their constant advice, collaboration and patience I owe a special gratitude to my colleagues at Bedford: Prof. H. O. W. Richardson, Prof. L. Pincherle; Drs. Nora Hill, R. Mansfield, M. R. Hoare, E. S. Owen Jones, N. M. Stewart, T. B. Taylor, R. N. Thomas, S. R. Mishra, I. A. Hassairi, Z. Aung, D. B. Rees, P. Pal and Joan Grace. For their excellent technical assistance I must thank Miss M. J. Franklin, Mrs. K. Wakeley, Miss S. Cohen, Miss S. McGrath; and Messrs. W. A. Baldock, F. Grimes, A. Betts, A. King, A. O. T. LeMottee, S. Bennett, J. Sales, D. Smee, Mrs. A. Gregory, M. Audria, M. Fesselet, M. Stürzinger, Mlle Lütke and Mme N. Ferrier. I am also grateful to the library staffs at both Bedford and CERN, who have been extraordinary helpful.

A scientist cannot work in this field of physics without institutional support, and I am pleased to acknowledge my debt to Bedford College, to CERN and to the Science Research Council.

August 1973　　　　　　　　　　　　　　　　Peter Rice-Evans

Contents

CHAPTER 1: PERTINENT PHYSICS

1.1 Introduction   19
1.2 Detection of Energetic-charged Particles   19
    1.2.1 Energy losses, relativistic rise   19
    1.2.2 Primary and secondary ionization   23
    1.2.3 Statistics of the energy loss by collision   25
1.3 Ionization Effects in Gases   29
    1.3.1 Excitation, ionization   29
    1.3.2 Recombination, light emission   33
    1.3.3 Negative ions, electron attachment   35
    1.3.4 Noble molecular ions   39
    1.3.5 Penning effect   41
    1.3.6 Charge exchange, quenching   42
1.4 Motion of Ions in the Electric Field   45
    1.4.1 Collisions, energy distributions   45
    1.4.2 Kinetic theory, drift and diffusion   47
    1.4.3 Growth of an avalanche, Townsend theory   55
    1.4.4 Growth of a streamer   60
    1.4.5 The Lozanskii theory of streamers   68
    1.4.6 Breakdown; the growth of a spark   74
    1.4.7 The statistics of avalanche growth   77
    1.4.8 Charge transport in noble liquids and solids   80

CHAPTER 2: HISTORICAL REVIEW

CHAPTER 3: NARROW-GAP SPARK CHAMBERS

3.1 An Elementary Spark Chamber   88
3.2 Rudiments   92
    3.2.1 Spark formation   92
    3.2.2 Techniques for construction   96
    3.2.3 Chamber gases   98

3.3	**Wire Planes**	100
	3.3.1 Principle	100
	3.3.2 Multiple sparks	100
	3.3.3 Attainable precision	103
	3.3.4 Construction	103
3.4	**Pulsing the Chambers**	104
	3.4.1 Requirements of a trigger system	104
	3.4.2 Production of the high-voltage pulse	106
	3.4.3 Application of the pulse to the chamber	112
	3.4.4 The clearing field	115
3.5	**Chamber Performance**	116
	3.5.1 Single-track efficiency	116
	3.5.2 Spatial resolution	118
	3.5.3 Multiple-track efficiency	119
	3.5.4 Time resolution	122
	3.5.5 Recovery time	124
	3.5.6 Performance in magnetic fields	125
	3.5.7 Performance at high pressures	127

CHAPTER 4: MEASUREMENT IN SAMPLING SPARK CHAMBERS

4.1	**Recording with Photographic Film**	129
	4.1.1 Optics	130
	4.1.2 Photography	134
	4.1.3 Digitization and analysis	134
	4.1.4 Precision	136
4.2	**Sonic Detection**	137
	4.2.1 Principles of measurement	137
	4.2.2 Examples of sonic chambers	138
	4.2.3 Advantages and difficulties	140
4.3	**Television Vidicon Camera Recording**	140
	4.3.1 Explanation of vidicon	141
	4.3.2 Application to spark chambers	142
	4.3.3 Accuracy	143
	4.3.4 Rosy prospects for vidicons	144

CONTENTS

4.4	**Magnetic Core Read-out**	145
	4.4.1 Ferrite cores	145
	4.4.2 Example of chambers	146
	4.4.3 Read-out	147
	4.4.4 Precision	148
	4.4.5 Difficulties	148
4.5	**Magnetostrictive Read-out**	149
	4.5.1 Magnetostrictive delay-line	149
	4.5.2 Application to spark chambers	151
	4.5.3 Digitization	152
	4.5.4 Precision	153
	4.5.5 Difficulties due to multiple-sparking and to magnetic fields	155
	4.5.6 Use of the torsional mode of magnetostriction	155
4.6	**Sparkostriction**	158
4.7	**Current Distribution Method**	160
4.8	**Capacitative and Other Memories**	163
4.9	**Vidicon Observation of Secondary Sparks**	165
4.10	**Inductive Location of Single Sparks**	167
4.11	**Assorted Methods of Recording**	168
	4.11.1 The delay-line chamber	168
	4.11.2 Digitizing with photodiodes	168
	4.11.3 Direct recording with magnetic tape	170

CHAPTER 5: WIDE-GAP SPARK CHAMBERS

5.1	**Introduction**	171
5.2	**Creation of the Spark Channel**	172
5.3	**High-voltage Pulse Generation**	176
	5.3.1 The Marx generator	176
	5.3.2 Miscellaneous HV generators	182
5.4	**Examples of Wide-gap Spark Chambers**	184
	5.4.1 Harvard chamber in a magnetic field	184
	5.4.2 Digitized chamber at CERN	185

12 CONTENTS

5.5	**Performance**	187
	5.5.1 Response to inclined tracks	187
	5.5.2 On gases and track quality	189
	5.5.3 Recording simultaneous tracks	190
	5.5.4 Memory and recovery times	192
	5.5.5 Precision of measurement	194
	5.5.6 Ionization effects	194

Chapter 6: Streamer Chambers

6.1	**Introduction**	196
	6.1.1 Preliminary	196
	6.1.2 Isolated streamers—side view	198
	6.1.3 Isolated streamers—end view	198
	6.1.4 Merged streamer regime	200
	6.1.5 General picture	201
	6.1.6 Aspects of timing	202
6.2	**High-voltage Pulse Shaping**	203
	6.2.1 Series and shunt spark gaps	204
	6.2.2 Blumlein line	206
	6.2.3 Measurement of high-voltage pulses	212
6.3	**Examples of Streamer Chambers**	215
	6.3.1 Bedford College streamer chamber	215
	6.3.2 The SLAC streamer chamber	219
	6.3.3 The CERN avalanche chamber	221
	6.3.4 The DESY (Hamburg) streamer chamber	222
6.4	**Recording in Streamer Chambers**	224
	6.4.1 Brightness of streamers	224
	6.4.2 Photographing streamers	227
	6.4.3 Image intensification	229
6.5	**Performance of Streamer Chambers**	230
	6.5.1 Precision of measurement	230
	6.5.2 Efficiencies and memory times of streamer chambers	232
	6.5.3 Performance with helium	234
	6.5.4 Performance with hydrogen	235

6.6	**Ionization Measurement in Streamer Chambers**	237
	6.6.1 Prospects for measurement	237
	6.6.2 Streamer brightness at Erevan	238
	6.6.3 Primary ionization in Moscow	238
	6.6.4 Ionization matters at CERN	242

CHAPTER 7: PROPORTIONAL CHAMBERS

7.1	**Introduction**	245
7.2	**Properties of Cylindrical Counters**	247
	7.2.1 Pulse formation in a conventional counter	247
	7.2.2 Amplification in a cylindrical geometry	252
	7.2.3 Fluctuations in pulse height	254
7.3	**Properties of Multiwire Chambers**	259
	7.3.1 The electric field	259
	7.3.2 Amplification and pulse formation	262
7.4	**Operation, Construction, and Related Problems**	265
	7.4.1 The Charpak prototype proportional chamber	265
	7.4.2 Large chambers, electrostatic problems	267
	7.4.3 Gas mixtures	269
	7.4.4 Constructional details	273
7.5	**Performance of Multiwire Chambers**	275
	7.5.1 Particle efficiency	275
	7.5.2 Time resolution and dead time	277
	7.5.3 Spatial accuracy	280
	7.5.4 Response in a magnetic field	281
	7.5.5 Performance at low pressures	282
	7.5.6 Operation in the Geiger-Müller mode	282
	7.5.7 Chamber cascades, ionization, and particle identification	284
7.6	**Electronic Amplifiers and Read-out**	293
	7.6.1 Pulse amplification and preparation	293
	7.6.2 Extraction of data	296
	7.6.3 Read-out employing induced positive pulses	298

Chapter 8: A Miscellany of Chambers

- 8.1 **Wide-gap Chambers** — 301
  - 8.1.1 Prospects for a laser-excited chamber — 301
  - 8.1.2 Microwave discharge chamber — 303
  - 8.1.3 Chamber with a rotating electric field — 305
- 8.2 **Hybrid Proportional/Spark Chamber** — 307
- 8.3 **Neon Flash Tubes** — 308
- 8.4 **Hybrid-expansion Chambers** — 311
  - 8.4.1 Streamer-cloud chamber — 311
  - 8.4.2 Towards the triggered bubble chamber — 312
- 8.5 **Noble Liquid Chambers** — 317
  - 8.5.1 Prospects for liquids — 317
  - 8.5.2 Electron multiplication in liquid argon — 318
  - 8.5.3 Experience with liquid xenon — 319
- 8.6 **Spark Counters** — 322
- 8.7 **Outwardly-directed Avalanche Chamber** — 323
- 8.8 **Stacks of Transparent Discharge Chambers** — 326

Chapter 9: Examples of Applications of Chambers

- 9.1 **Particle Physics** — 328
  - 9.1.1 Observation of two kinds of neutrino — 328
  - 9.1.2 Splitting the A_2 resonance at CERN — 329
  - 9.1.3 Colliding protons at CERN — 330
  - 9.1.4 Photoproduction of hadrons at DESY — 331
  - 9.1.5 Transition radiation detection — 332
  - 9.1.6 Search for fractionally charged particles — 335
- 9.2 **Low-energy Physics** — 338
  - 9.2.1 Alpha rays in a streamer chamber — 338
  - 9.2.2 Rare events—search for double beta decay — 341
  - 9.2.3 Fermi surfaces—positron annihilation photons — 342
- 9.3 **Cosmic Rays** — 344
  - 9.3.1 Spark chambers for cosmic rays — 344
  - 9.3.2 Gamma-ray astronomy — 345

CONTENTS 15

9.4	**Study of Electrical Discharges**	347
	9.4.1 Electrical breakdown along an ionized trail	347
	9.4.2 Streamer formation in hydrogen	348
	9.4.3 The ringing of Lichtenberg figures	350
9.5	**Applications in Medicine and Biology**	352
	9.5.1 Applications of isotopes in medical diagnosis	352
	9.5.2 The spark chamber approach	353
	9.5.3 Proportional chambers for nuclear medicine	356
9.6	**Archaeology**	357
	9.6.1 Chephren's pyramid	357

CHAPTER 10: DRIFT CHAMBERS

10.1	**The Basic Idea**	360
10.2	**The Choice of Gas**	361
	10.2.1 Drift velocity	361
	10.2.2 Electron diffusion	363
10.3	**Examples of Drift Chambers**	365
	10.3.1 Developments at CERN	365
	10.3.2 The Saclay chamber	368
	10.3.3 The Heidelberg chamber	369
	10.3.4 The giant Harvard chambers	370
10.4	**Read-out Electronics**	373
	10.4.1 Introduction	373
	10.4.2 The digital approach	373
	10.4.3 The analogue method	375
	10.4.4 Time stretching	375
	10.4.5 Reading the second coordinate	376
10.5	**Performance**	378
	10.5.1 Accuracy	378
	10.5.2 Angled tracks and left-right ambiguities	379
	10.5.3 Time resolution and multiple tracks	380
	10.5.4 Use in magnetic fields	381

Appendix A: High-voltage Properties of Materials   383
Appendix B: Preparation of Plastic Scintillators   384
Bibliography   385
Index   401

Frontispiece: Dr Peter Schübelin inspecting his wide-gap spark chambers at CERN. (photo CERN)

Σὰ βγεῖς στὸν πηγαιμὸ γιὰ τὴν Ἰθάκη,
νὰ εὔχεσαι νἆναι μακρύς ὁ δρόμος.

C. P. Cavafy

CHAPTER 1

Pertinent Physics

1.1 INTRODUCTION

IN this chapter we shall present various aspects of physics that are relevant to spark chambers. The volume of work that has been reported in the fields of ion and gas-discharge physics is considerable. At first glance it appears as a mass of uncorrelated and sometimes contradictory data. The difficulties associated with experimentation—e.g. the control of gas purity—and the statistical nature of many of the phenomena have meant that empirical formulae have often been developed, rather than rigorous theories. For many purposes these formulae are very satisfactory—but it must always be remembered that they are often valid only over a limited range of conditions.

1.2 DETECTION OF ENERGETIC-CHARGED PARTICLES

1.2.1 Energy losses, relativistic rise

When an energetic-charged particle passes through a gas it undergoes a series of inelastic Coulomb collisions with the electrons in the gas. As a result, it loses energy by excitation and ionization of the gas molecules, and this can be used to display the line of the trajectory. For the purposes of this monograph other effects, such as the occasional interaction with nuclei, bremsstrahlung losses, and Čerenkov radiation, will be ignored.

It may be noted that the energy imparted to an electron on ionization will depend on the collision distance. Knock-on electrons (delta rays) result from larger energy transfers and correspond to close collisions. However, the majority of the free electrons come from more distant collisions and have small kinetic energies.

The theory for ionization loss has been given by Bethe (see Ritson, 1961; Rossi, 1965). Two cases must be distinguished:

(*a*) heavy particles (muons, protons, etc.), and (*b*) electrons and positrons. In the first case the rate of energy loss is given by:

$$-\frac{dE}{dx} = \frac{2Dm_ec^2z^2}{\beta^2}\left[\ln\frac{4m_e^2c^4\beta^4}{(1-\beta^2)^2 I^2(Z)} - 2\beta^2\right],$$

where βc is the particle velocity, m_e the electron mass, z the particle charge, and

$$D = \pi N \frac{Z}{A} r_e^2 = 0.150 \frac{Z}{A} \text{ g}^{-1} \text{ cm}^2,$$

where Z and A are the atomic and mass numbers of the gas atoms, N is Avogadro's number, and r^e is the classical radius of the electron (e^2/m_ec^2).

In the case of electrons:

$$-\frac{dE}{dx} = 2Dm_ec^2z^2\left[\ln\frac{\pi^2(m_ec^2)^2}{(1-\beta^2)^{3/2} I^2(Z)} - a\right],$$

where $a = 2.9$ for electrons and 3.6 for positrons.

The factor D expresses the proportionality of the collision probability to the electron density. The term $I(Z)$ represents the geometric mean of all the ionization and excitation potentials of the gas, and was first suggested by Bloch to be

$$I(Z) = I_H Z,$$

where $I_H = 13.5$ eV is the energy corresponding to the Rydberg frequency. More recently, Sternheimer (1966) has expressed $I(Z)$ by the empirical relation (for $Z \geqslant 13$)

$$I = Z(9.76 + 58.8 Z^{-1.19}) \text{ eV},$$

see column D, Table 1.2.2.

It is clear that for particles of the same charge, the energy loss is a function only of β. When particles of various energies are considered, the smooth curves of Fig. 1.2.1.1 are obtained. A feature to be noted is the initial decrease with rising energy until the range of minimum ionizing power is reached. Thereafter, the relativistic $(1 - \beta^2)$ term in the logarithm causes a rise in the curve.

However, the rise does not persist, as might be expected, because the dielectric properties of the medium reduce the field at large impact parameters and consequently the curve levels

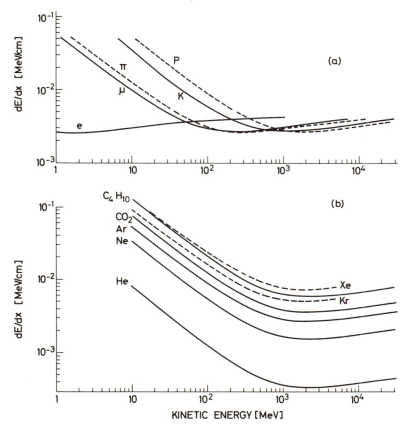

Fig. 1.2.1.1 (a) Energy loss for various particles in argon (density 0·00178 gm cm^{-3}); (b) Energy loss for protons in various gases. (Dimčovski, 1970)

out to a plateau (Fig. 1.2.1.2). The ratio of the plateau/minimum ionizing power, which is important in a number of experiments, is heavily dependent on the nature of the gas, and its density. Calculations have been made by Budini *et al.* (1960) and Ermilova *et al.* (1969), and although some ambiguity remains depending on the choice of I and also on what is considered the maximum transferable energy in a collision, there can be good agreement between theory and experiment. Actually, a good comparison has related the primary specific ionization (see next

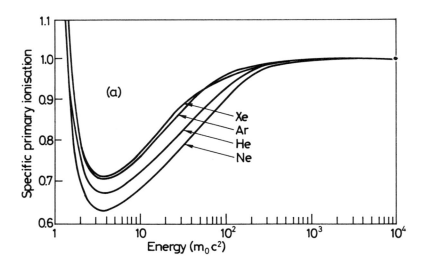

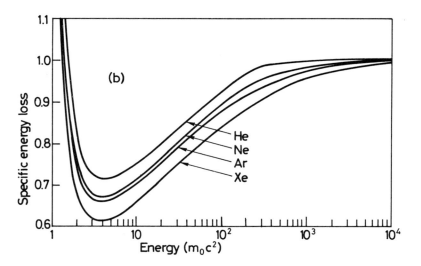

Fig. 1.2.1.2 Relativistic growth with particle energy of (a) specific primary ionization; (b) specific energy loss in noble gases. The ordinates are expressed as ratios of the plateau value, and the energy is in units of rest mass. (Ermilova *et al.*, 1969)

section—1.2.2) with the recent streamer chamber measurements on helium (Ermilova *et al.*, 1969). The rise to a plateau is known as the density effect (see section 6.6.3). Calculated values for the relativistic rise in the noble gases at atmospheric pressure are given in Table 1.2.2. They include 48·7% for helium and 58·7% for neon; but it must be remembered that the values would be considerably higher for lower pressures and vice versa. Mulvey (1973) has suggested that the rise in xenon could be as high as 90%.

It has been postulated that the balance of the energy loss predicted goes off in Čerenkov radiation.

1.2.2 Primary and secondary ionization

We have seen that a particle traversing a gas loses energy by excitation and ionization of the gas molecules. A proportion of the electrons ejected in the ionization process will gain kinetic energies large enough to cause further, secondary ionization. It is, therefore, important to distinguish between the number of ionizing collisions engaged in by the particle per unit length (*primary specific ionization, j (primary)*) and the total number of ion pairs that are actually created in unit path length (*total specific ionization, j (total)*). Of course, these are averages.

The values, corresponding to minimum ionization, for the two classes of specific ionization, are shown in columns F and I, Table 1.2.2. For example, in neon a particle will make about twelve collisions per cm, but a total of thirty-seven electrons will be liberated. The thirty-seven electrons will initially be distributed near the collision sites, but subsequently they will drift apart. The numbers should be treated with some reserve; it is notoriously difficult to measure the quantities accurately. The approximate factor of 3:1 is, however, a useful guide.

Rossi (1965) gives the following expression for the specific primary ionization:

$$j_{\text{primary}} = \frac{2Dm_e c^2}{\beta^2} z^2 \frac{r}{I_0} \left[\ln \frac{2m_e c^2 \beta^2}{(1-\beta^2)I_0} + s - \beta^2 \right],$$

where r and s are dimensionless constants, D is again related to the electron density of the gas and I_0 is the ionization potential of the outer shell of the atom.

TABLE 1.2.2

Ionization parameters for gases at atmospheric pressure and 0°C, culled largely from Ermilova et al. (1969). Column E is from Jesse and Sadauskis (1955); column F from Charpak et al. (1965); and column D calculated from Sternheimer (1966). Many of the results in column I are ancient; the best experiment is that with a streamer chamber on helium (Davidenko et al., 1969).

A	B	C	D	E	F	G	H	I	J
						Primary specific ionization (cm^{-1})			
Gas	Gas density $\rho \times 10^3$ g/cm^3	Ionization potential eV	Average ionization potential $I(Z)$ eV	Average energy to create one ion pair (V_0) eV	Total specific ionization cm^{-1}	Plateau (calc.) cm^{-1}	Min. ion. (calc.) cm^{-1}	Min. ion. (expt.) cm^{-1}	Relativistic increase in primary specific ionization %
He	0·17847	24·56	43·3	42·3	~11	5·2	3·5	3·83	48·7
Ne	0·90035	21·66	141·4	36·6	~37	18·1	11·4	12·4	58·7
A	1·78370	15·75	203·5	26·4	~110	36·5	25·8	27·8	41·6
Xe	5·8510	12·08	527·8	22·2		69·6	49·6	44	40·6
H$_2$	0·08988	15·4	19·9	36·3		6·9	5·1	5·32	36·1
N$_2$	1·25055	15·5	94·8	34·7		39·0	27·1	19·3	43·8
O$_2$	1·42904	12·2	97·3	30·9		44·0	28·9	22·2	52·2
Air				33·9			27·5	18·5	

In connection with the total ionization, a parameter for a particular gas is the average energy V_0 lost by the incident particle in the creation of every ion pair. It is a useful parameter because its value appears to be relatively insensitive to the type and energy of the particle (see column E, Table 1.2.2) and because it is simply related to the total specific ionization:

$$j_{\text{total}} = \frac{dE/dx}{V_0}.$$

When the incident particle engages in a close collision, the atomic electron, having considerable kinetic energy, then careers off causing a trail of ions that is distinguishable from the ionized path of the particle. A study of these knock-on electrons (delta rays) can give information similar to that obtained in ionization measurements. Ritson (1961) shows that the probability of a close collision that imparts an energy E is given by

$$P(E)\,dE = W\frac{dE}{E^2},$$

where

$$W = \frac{0\cdot 15}{\beta^2}\frac{Z}{A}z^2 x \text{ MeV},$$

and where x is the thickness of matter traversed in g/cm². This shows that the number of delta rays with energies greater than a chosen value E, in an average track is just

$$\int_E^\infty \frac{W\,dE}{E^2} = \frac{W}{E},$$

i.e. a count will give a measure of β (or z for $\beta = 1$). Ritson gives the example of a particle, with $\beta = 1$, traversing 1 g/cm² of material with $Z/A = \frac{1}{2}$. Thus $W = 75$ kV, and one would expect to find five delta rays with energies exceeding 15 kV. For a relativistic track with $\beta = 1$, the density of delta rays is just a function of z^2 where z is the charge of the incident particle.

1.2.3 Statistics of the energy loss by collision

In passing through matter, the collisions made by a charged particle are with the atomic electrons. Each collision will be an independent interaction and will therefore be statistical in nature. In a "thin absorber" such as a gas (i.e. an absorber in which the ionization losses are much smaller than the particle energy), different statistics apply depending on whether primary or total ionization is determined.

The primary ionization indicates the actual number of collisions (N) that have occurred. For identical particles crossing a given length of gas, this number will not fluctuate much, and the situation can be described by the Poisson distribution. The r.m.s. fluctuation will be given by $N^{\frac{1}{2}}$, and the accuracy of measurement may be improved by increasing the track length and hence increasing N.

On the other hand, the value for the total ionization includes the ions produced by the occasional delta rays, and this number is subject to wild fluctuations from one trail to another. In this case, doubling the track length merely doubles the chance of obtaining a large fluctuation; and the accuracy of measurement is hardly improved. The statistics necessary to tackle this case have been derived by Landau (1944), Symons (1948), Vavilov (1957) and Blunck and Liesegang (1950).

Rossi (1965) has summarized the Landau statistics and has given the procedure, with graphs of the necessary parameters, to enable one to plot specific distributions. The significant characteristic is the pronounced tail at large ionizations in the Landau distribution. This is seen in Fig. 1.2.3.1, where the results are shown of a streamer chamber experiment on primary and total ionization in helium (Davidenko, Dolgoshein et al., 1969). The agreement between the results and the statistics enabled the authors to use the measured distribution as a check on the type of ionization determined. From the point of view of measuring ionizing power, it is clear that advantage lies in using a regime that is described by the sharper Poisson distribution.

Whereas Landau had assumed that any energy up to infinity could be transferred by a particle to a free electron, Vavilov (1957) pointed out that one should impose an upper limit to the transfer:

$$\varepsilon_{\max} = \frac{2m_e c^2 \beta^2}{1 - \beta^2} \left[1 + \frac{2m_e}{M} \frac{1}{\sqrt{(1 - \beta^2)}} + \left(\frac{m_e}{M}\right)^2 \right]^{-1}.$$

STATISTICS OF THE ENERGY LOSS BY COLLISION

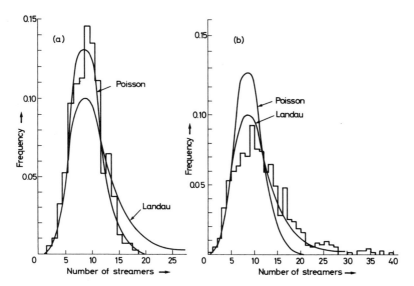

Fig. 1.2.3.1 The Poisson and Landau distributions fitted to streamer chamber measurements of ionization in helium at 0·4 atmospheres. The delay before application of the HV pulse was 200 nsec in (a) and 15 μsec in (b). The histogram and curves have been normalized on the basis of area. The ordinate is the probability of obtaining a given number of streamers in a track length of 5 cm. (Davidenko et al., 1969)

The effect of this modification is to sharpen the Landau curve and particularly to diminish the tail at large energy losses.

This, however, is not the end of the story; for the Landau distribution is limited in its application. This is because its derivation ignores the fact that the electrons are bound to the atom. As Rossi remarks, the Landau distribution is not valid for "extremely thin absorbers" for which the electrons, involved in distant collisions, cannot be considered free.

Blunck with Liesegang (1950) and Westphal (1951) have derived the distribution for the case that includes electron binding effects. The distribution has been supported experimentally by Ramana Murthy and Demeester (1967) using proportional counters. To illuminate the matter we shall outline their discussion in what follows.

They found they were able to represent the Landau distribution with the empirical formulae, of the type

$$F(\Lambda)\,d\Lambda = 0{\cdot}762(2\pi)^{-\frac{1}{2}} \exp\{-0{\cdot}5(\Lambda + e^{-\Lambda})\}\,d\Lambda$$
$$\text{for } \Lambda \leq 0$$
$$= 0{\cdot}762(2\pi)^{-\frac{1}{2}} \exp\{-0{\cdot}5(\Lambda^{0{\cdot}85} + e^{-\Lambda})\}\,d\Lambda$$
$$\text{for } 0 < \Lambda < 10.$$

Here Λ is a dimensionless parameter and is related to any arbitrary energy loss (Δ) in the counter by

$$\Delta = \Delta_{\text{most probable}} + \Lambda\{0{\cdot}300mc^2(Z/A)x\},$$

where m = rest mass of electron
Z = atomic number of the medium
A = atomic weight of the medium
x = thickness of the gaseous medium in g cm^{-2}.

They also found that the Blunck and Liesegang distribution could, in their experimental conditions, be represented by the empirical formulae

$$F(\Lambda)\,d\Lambda = 0{\cdot}088 \exp(-\Lambda^2/26), \text{ for } \Lambda \leq 0$$
$$= 0{\cdot}145 \exp[-0{\cdot}5\{\tfrac{1}{2}\Lambda + \exp(-\tfrac{1}{2}\Lambda)\}]$$
$$\text{for } 0 < \Lambda < 12$$

where Λ has the same meaning as above.

The question, which distribution to use, is answered by referral to a parameter b defined such that

$$b^2 = \frac{\bar{\Delta}(\text{eV})Z^{4/3} \times 20(\text{eV})}{\xi^2(\text{eV})^2}$$

where $\bar{\Delta}$ = average energy loss in traversing the counter of
x g cm^{-2}
$\xi = 0{\cdot}300(Z/A)x(m_e c^2/\beta^2)$
m_e = mass of the electron
βc = velocity of the incident particle.

Blunck and Liesegang suggest that when $b^2 \ll 3$ no correction to the original Landau distribution is required; but when $b^2 \gg 3$ their own much wider distribution should be used.

In their investigations with a 15 cm deep proportional counter [argon + methane (93:7), 2370 volts, multiplication factor 75],

EXCITATION, IONIZATION

Ramana Murthy and Demeester first studied the energy loss distribution suffered by a beam of 80 MeV protons. The calculated value of b^2 was 1·184 and there was excellent agreement between the experimental data and both the Landau and Blunck and Liesegang distributions. On the other hand, when the counter was subjected to a 4·0 GeV π^- beam the distribution shown in Fig. 1.2.3.2 was obtained. The corresponding value of b^2 was 13 and it is clear that the data are best fitted with the Blunck and Liesegang curve.

An excellent introduction to the use of the two distributions, including a lot of computational detail, has been given by Dimčovski (1970). However, the situation is still not satisfactory. As will be seen in section 7.5.7, curves obtained with stacks of Charpak chambers lie between the theoretical curves discussed here.

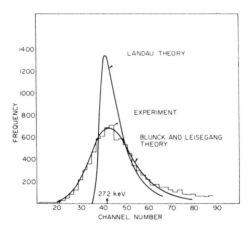

Fig. 1.2.3.2 Comparison of theory with the observed distribution of energy losses suffered by 4·0 GeV negative pions, in a path length of 15 cm in a proportional counter. (Ramana Murthy and Demeester, 1967)

1.3 IONIZATION EFFECTS IN GASES

1.3.1 Excitation, ionization

After the passage of a charged particle, a trail of ions, electrons and excited atoms lies in its wake. In the spark chamber, partly under the influence of the electric field, they will interact. Although the detail is often uncertain, we shall outline some of the main processes.

Atoms may be raised to discrete excited states by electron bombardment:

$$X + e \to X^* + e.$$

The minimum energy required for this process is the excitation energy of the specific level. However, the laws of conservation and quantization of angular momentum impose restrictions on the possible collisions, and consequently the probability for excitation of a particular level, as a function of electron energy, may take the form shown in Fig. 1.3.1.1. It is seen that the curves rise steeply to a maximum, and typical values for the cross-section lie in the region 10^{-17} cm^2.

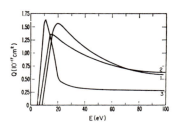

Fig. 1.3.1.1 Measured cross-sections for excitation of the 2p$_8$ levels by electron impact. The curves are for (1) argon, (2) krypton and (3) xenon. (Moiserwitsch and Smith, 1968)

Ionization, i.e. the release of a free electron from a neutral atom or molecule, may be caused by collision with an energetic electron:

$$X + e \to X^+ + e + e.$$

The conditions concerning angular momentum are not so strict for ionization because a three-body system results from the collision. The minimum energy for the process is given by the ionization potential. In Fig. 1.3.1.2 it may be noted that the screening effect of the orbital electrons is such that the potentials

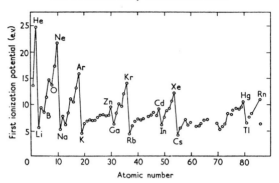

Fig. 1.3.1.2 Variation of first ionization potentials with atomic number.

rise as the electron shell is filled, culminating in the noble gases. Table 1.3.2 includes some first and second ionization and excitation potentials. A larger energy is required for the second ionization, etc.

The variation of ionization cross-section for noble gases with electron impact energy is given in Fig. 1.3.1.3. Typical values are more than 10^{-16} cm, i.e. higher than the excitation cross-sections, but, of course, their threshold is also higher. The cross-section is defined in such a way that the value at a particular electron impact energy refers to all ejected electrons—regardless of their final kinetic energy. In general, most of the electrons are emitted with small energies, which means that the bombarding electron loses not much more than the energy corresponding to the ionization potential.

Excitation and ionization of atoms and molecules may also be caused by collisions with photons. In the case of excitation, the photon energy must correspond with the discrete energies of the excited states:

$$h\nu + X \to X^*.$$

Even when the photon energy is correct, the operation of selection rules means that certain levels will be strongly excited, others not. If the photon energy be greater than the ionization potential, photoionization may occur:

$$h\nu + X \to X^+ + e.$$

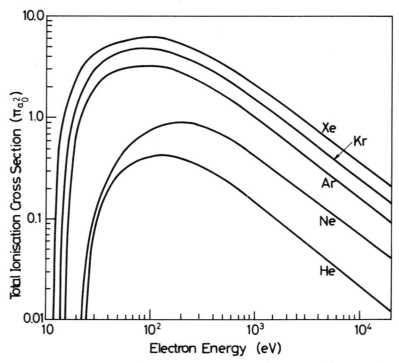

Fig. 1.3.1.3 Total ionization cross-sections for the noble gases by electron impact ($\pi a_0^2 = 8.8 \times 10^{-17}$ cm^2). (Rapp and Englander-Golden, 1965)

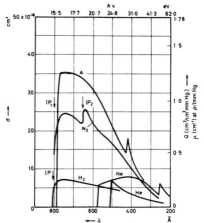

Fig. 1.3.1.4 Photoionization cross-sections for various gases. (Francis, 1960)

Any excess energy is taken away as kinetic energy. A graph showing the cross-sections for photoionization in the noble gases is given in Fig. 1.3.1.4. It is noted that the curves show a sharp rise from the ionization energy, but that the absolute magnitudes are not as large as for ionization by electrons.

1.3.2 Recombination, light emission

In a gas containing positively and negatively charged particles, radiative recombination is likely to occur:

$$X^+ + Y^- \rightarrow XY + h\nu$$

and

$$X^+ + e \rightarrow X + h\nu.$$

In this process the photon has to take away the energy liberated in the recombination plus the kinetic energy of the electron. Thus the photon energy spectrum will reflect the electron velocity distribution which may well be continuous. In the recombination of doubly ionized atoms

$$X^{++} + e \rightarrow X^+ + h\nu,$$

the photon may be capable of causing photoionization in the gas because of the higher value for the second ionization coefficient. Other modes of recombination are

$$X^+ + Y^- \rightarrow X + Y$$
$$X^+ + Y^- + Z \rightarrow X + Y + Z$$
$$XY^+ + e \rightarrow X + Y$$

and the conservation laws must apply.

The concentration of ions (n_+ and n_-) at any moment, if they be uniformly distributed in a gas, may be expressed in terms of the recombination coefficient (β) (Meek and Craggs, 1953). This is defined such that the number of recombinations during dt is

$$dn = -\beta n_+ n_- \, dt$$

and with dimensions

$$\beta = vQ,$$

where v, for electron recombination, is the electron velocity and

Q the cross-section. If $n_- + n_+ = n$, we have the Thomson–Rutherford (1896) recombination law

$$\frac{dn}{dt} = -\beta n^2.$$

If n_0 is the concentration at $t = 0$, we obtain

$$n = \frac{n_0}{1 + \beta n_0 t}.$$

The decay is thus inversely proportional to time (and not exponential). Experimental values for the recombination coefficients differ considerably, but some approximate values are given in Table 1.3.2. One thing to notice is that the rate in rare gases increases with atomic number. It appears that recombination is exceedingly complicated, and the reader is recommended to consult Hasted (1964) for a detailed discussion.

In general, the radiation from a discharge or spark is a function of the spectrum of the gaseous elements and the temperature. As we have observed the influence of the electron velocities is likely to transform the line spectrum into a broad continuum about the lines. An approximate knowledge of a particular discharge spectrum is sometimes useful if one wishes to optimize the conditions for photography.

The most important spark chamber gas is neon. Its energy level diagram is shown in Fig. 1.3.2.1. Most of the excited states have lifetimes of the order of 10^{-8} sec but two ($2s_5$ and $2s_3$) are metastable, with lifetimes of 0·1 sec or more. Transitions to the ground state are in the ultra-violet, but those from the ten p-levels ($2p_1$–$2p_{10}$) to the four s-levels ($2s_2$–$2s_5$) cause the emission of a number of visible spectral lines, which together give the neon discharge its characteristic red colour.

Wavelengths, for what the *Handbook of Chemistry and Physics* refers to as "persistent emission lines", are recorded in Table 1.3.2 for some relevant gases. They are only a very rough guide however; for a proper list in the same volume mentions thirty lines for argon just in the wavelength range 4000–5000 Å.

An example of the importance of knowing the spectral emission is the use of hydrogen as the main gas in a streamer chamber. Corrigan and von Engel (1958) showed that less than 20% of radiation from a discharge in hydrogen has a wavelength

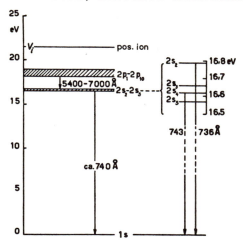

Fig. 1.3.2.1 Energy levels in neon. The 2s levels are expanded on the right. (Penning, 1957)

above 1800 Å, and so one may conclude that only a small proportion is emitted in the visible region. Thus in a hydrogen chamber, most of the radiation from the streamers would go unused by normal photographic film. To enhance the visible emission, the possibility exists of adding a small quantity of impurity gas (perhaps methane) to shift the wavelength of the ultra-violet (Grey Morgan, 1969).

Finally, on the subject of emission, Frankel et al. (1966) observed that when a helium spark chamber (at 1 atmosphere pressure) was pulsed, the radiation produced spurious tracks in a second chamber. After a few tests they concluded that X-rays with energies greater than 12 keV were emitted. No explanation was attempted, and neither were X-rays found when neon was substituted for the helium.

1.3.3 Negative ions, electron attachment

It is well known that in some gases, atoms exist largely in the molecular state. This is true of hydrogen, oxygen, nitrogen, etc., but not of the noble gases. The binding energy of H_2 is 4·48 eV and O_2, 5·1 eV.

The energy required to create an ion from a simple atom is

TABLE 1.3.2

Ionization and excitation data for a number of gases. Column E is from Meek and Craggs (1953). Column F, giving approximate experimental values, is from Hasted (1964). Column G lists the persistent lines and is from the *Handbook of Chemistry and Physics*, Chemical Rubber Co., Cleveland, Ohio.

A	B	C	D	E	F	G
Gas	Atomic number	First ionization potential (eV)	Second ionization potential (eV)	First excited state (eV)	Recombination coefficient ($cm^3 s^{-1}$)	Principal emission wavelength (Å)
He	2	24·48	54·40	20·9 19·8 meta	1×10^{-8}	584 3888 5875
Ne	10	21·56	41·07	16·68 16·53 meta 16·62 meta	2×10^{-7}	734 743 5400 5832 5852 6402
Ar	18	15·76	27·62	11·56 11·49 meta 11·66 meta	9×10^{-7}	1048 1066 6965 7067 7503 8115
Kr	36	14·00	24·56	9·98 9·86 meta 10·51 meta	1×10^{-5}	1236 5570 5870
Xe	54	12·13	21·2	8·39 8·28 meta 9·4 meta	2×10^{-6}	1296 1470 4501 4624 4671
H	1	13·60		10·2		1215 4861 6562
N	7	14·53	29·59	6·3		1200 4110
O	8	13·61	35·11	9·1		1302 7771
H_2		15·4		11·2	2×10^{-6}	
N_2		15·8		6·1	2×10^{-6}	
O_2		12·5			4×10^{-7}	
I_2		9·0		1·9	$1·5 \times 10^{-8}$	1782 2062

just the ionization potential (see Table 1.2.2). Similarly, an electron may be removed from a molecule to leave a positive molecular ion, e.g. under electron bombardment:

$$e + O_2 \to O_2^+ + 2e.$$

The energy required for this process in H_2, N_2, O_2 is given in Table 1.3.2. Both atomic and molecular ions may themselves be excited or further ionized.

Another class of ion is the heavy negative ion—in which an electron has attached itself to the neutral atom or molecule. Such a system is stable for many substances; especially those whose outer electronic shells are almost full. The energy released in the creation of a negative ion is called the electron affinity. This energy varies from about 4 volts (for gases like F) to nearly zero for those gases which do attach electrons, and is negative for those which do not (He, Ne, etc.) (for values see Brown, 1959). (It was ignorance of the role of electron attachment in oxygen, i.e. the absorption of the free electrons, that delayed an earlier development of the spark chamber.)

Negative ions may be created by a number of processes, e.g. radiative attachment (Branscomb, 1962):

$$e + X \to X^- + h\nu;$$

by dissociative attachment (Prasad and Craggs, 1962):

$$e + XY \to XY^- \to X + Y^-;$$

by ion pair formation:

$$e + XY \to X^+ + Y^- + e;$$

and by three-body collision

$$e + Ne + O_2 \to O_2^- + Ne.$$

The second process is essentially a resonance capture—the electron must have the correct energy to produce XY^- on a suitable potential curve. A typical example of a cross-section curve is that for O_2 given in Fig. 1.3.3.1

$$e + O_2 \to O^- + O,$$

where 6·5 eV is required. A very electronegative gas such as SF_6 requires practically no electron energy for negative ion production. The ion pair formation process requires more energy—approximately the pertinent ionization energy to release the electron (Fox, 1957).

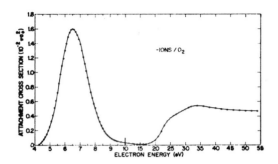

Fig. 1.3.3.1 Measured cross-sections for negative-ion formation in O_2 by electron impact. (Rapp and Briglia, 1965)

The attenuation of an electron swarm current in drifting through a gas may be described by

$$dI_e = -I_e \eta_a \, dx,$$

where the "attachment coefficient" η_a is the probability of electron attachment to a gas molecule in travelling one cm. Whence

$$I_e = I_{e0} e^{-\eta_a x}.$$

The coefficient is related to the cross-section for attachment by

$$\eta_a = \frac{n_0 \sigma_a \bar{v}_e}{u_e},$$

where $\bar{v}_e$ and u_e are the thermal and drift velocities for the electrons and n_0 the concentration (see later section).

Attachments of interest in spark-chamber work include water

$$e + H_2O \rightarrow H_2 + O^-$$

and alcohol

$$e + C_2H_5OH \rightarrow C_2H_5O^- + H.$$

Also of importance in spark chambers are the well-known fluor- and chlor-carbons, and related compounds such as SF_6, which have high dielectric strengths. Warren et al. (1950) have suggested that their low Townsend ionization coefficients are to be explained by dissociation reactions of the form

$$XY_n + e \to XY_n^+ + e + e$$
$$XY_n + e \to XY_{n-1}^+ + Y^- + e$$
$$XY_n + e \to XY_{n-1}^+ + Y + e + e,$$

where X is usually an atom of carbon or sulphur and Y is a halogen atom, usually chlorine or fluorine. In their experiments on $CClF_3$, CCl_2F_2 (freon-12), and CCl_3F, Warren et al. found that the second and third reactions were dominant and that the escaping atom Y was chlorine.

Some electron affinities of atoms forming stable negative ions are 0·8 eV for H, 1·5 eV for O, 3·6 eV for F, 3·7 eV for Cl, and 3·10 eV for I.

Some (approximate) experimental values for the maximum attachment cross-sections of various gases are given in Table 1.3.3 together with the energy of the bombarding electron at which the maximum occurred (Hasted, 1964).

TABLE 1.3.3

Gas	O_2	CO_2	SF_6	CCl_4	CCl_2F_2	H_2O
σ_{max} (10^{-16} cm²)	0·013	0·005	5·7	1·3	0·54	0·48
Te_{max} (eV)	6·2	7·8	0·00	0·02	0·015	6·4

1.3.4 Noble molecular ions

Although the noble gases do not normally form neutral diatomic molecules, the molecular ions He_2^+, Ne_2^+, Ar_2^+, Kr_2^+ and Xe_2^+ have been observed (Hornbeck and Molnar, 1951). It is suggested they are created in the process (see Fig. 1.3.4.1)

$$He + e \to He^* + e$$
$$He^* + He \to He_2^+ + e.$$

The electron energy required to cause this sequence appears to be about one or two volts less than the atomic ionization potential. For He_2^+, Ne_2^+, Ar_2^+ and Kr_2^+, the appearance potentials are 23·18, 20·86, 15·06 and 13·23 volts, respectively (cf. the ionization potentials in Table 1.2.2).

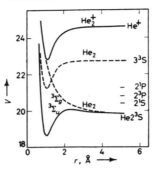

Fig. 1.3.4.1 Potential energy diagram for the helium molecule, illustrating the viability of the He$_2^+$ ion. (Hasted, 1963)

The ionization curve obtained with a mass spectrometer by Hornbeck and Molnar for helium is shown in Fig. 1.3.4.2. In their experiments, they noted that the ratio of peak heights of atomic to molecular ions was 85:1 in helium, 500:1 in neon, 1200:1 in argon, 20,000:1 in krypton and 40,000:1 in xenon. These measurements appear to be at pressures in the region 10^{-3} mm Hg, but it is pointed out that atomic ion production increases in proportion to the pressure (p), but molecular ion production with p^2. Fite et al. (1962) show that about 100 μsec

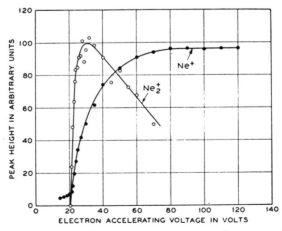

Fig. 1.3.4.2 Ionization in neon as a function of electron bombardment energy, showing the production of the Ne$_2^+$ ion. Similar curves obtained for Ar$_2^+$ and He$_2^+$. (Hornbeck and Molnar, 1951)

after a radio frequency discharge in helium at 2 Torr, the He_2^+ ions actually become more abundant than the He^+. Other possible production processes are (Lozanskii, 1969)

$$X^* + X^* \to X_2^+ + e,$$

but this is likely to be rare because of the low density of X^* to X. Also (Dolgoshein et al., 1969)

$$X^* + X + X \to X_2^+ + X + e$$

and

$$X^* + X + X \to X_2^+ + X.$$

This reaction is stated to occur within a time 10^{-8} sec at atmospheric pressure.

Composite ions have been considered (Kaul et al., 1963), e.g. $NeAr^+$

$$Ne^* + Ar \to NeAr^+ + e.$$

With regard to the dissociation of the molecular ion, Willis and Grey Morgan (1969) have referred to

$$X_2^+ \to X^+ + X + h\nu,$$

and Dolgoshein et al. (1969) to

$$X_2^+ + e \to X^* + X$$

and

$$X_2^+ + e + X \to X_2^* + X \quad \text{(in a dense gas)}$$

and

$$X_2^+ + e + e \to X_2^* + e \quad \text{(in a strongly ionized plasma)}.$$

We notice that excited neutral noble molecules are invoked in the last two cases.

1.3.5 Penning effect

As a rule, the de-excitation of an atom is a straightforward emission of a photon with a lifetime in the region of 10^{-8} sec. However, long-lived metastable states produced by electron collisions are an exception—for here, radiation to the ground state is forbidden by selection rules. De-excitation of metastable states may occur in collisions with other atoms: the surplus energy may excite the second atom or, in a special case of interest it may ionize the second atom (Penning effect). The necessary condition is that the energy of the metastable level must be greater than the ionization potential of the second collision particle.

The ionization produced by this effect can substantially increase the number of free electrons that participate in an electrical discharge and hence, in a spark chamber, facilitate spark growth. An example is the addition of a little argon to neon. As can be seen in Table 1.3.2, neon has a metastable level of 16·53 eV, which is greater than the ionization potential of argon (15·76 eV). Hence in a collision

$$Ne^* + Ar \rightarrow Ne + Ar^+ + e.$$

Another possibility is the addition of argon to helium. Hasted (1964) quotes the Penning ionization cross-sections for $(Ne^* + Ar)$ and $(He^* + Ar)$ as $2 \cdot 6 \times 10^{-16}$ and $7 \cdot 6 \times 10^{-16}$ cm^2, respectively. The Penning effect will also occur if molecular gases are added to noble gases—for the former have low ionization potentials.

The curves in Fig. 1.3.5.1 show the proportion of electron energy dissipated in elastic, exciting and ionizing collisions as a function of the field. The additional ionization obtained by adding a little argon to neon is clearly illustrated.

1.3.6 Charge exchange, quenching

A phenomenon that has important practical applications is charge exchange. When an ion collides with a molecule, an electron is likely to be exchanged if the ionization potential (I) of the ion is greater than that of the molecule. The cross-sections for the process can approach the kinetic theory collision cross-sections, especially when at least one of the bodies has a complicated structure. An application of charge exchange is the addition of ethyl alcohol to argon in the Geiger counter. The respective ionization potentials are 11·5 and 15·7 V, so that argon ions are soon neutralized and replaced by alcohol ions.

In many discharge devices, positive ions are eventually collected at a metallic cathode. When the ion has approached to

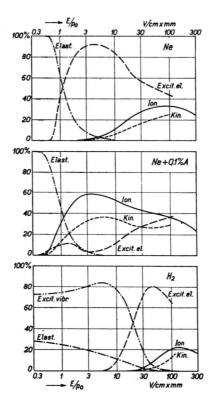

Fig. 1.3.5.1 Fraction of electron energy (acquired from the electric field) that is spent in elastic collisions (Elast.), excitation of electronic levels (Excit. el), excitation of vibrational levels (Excit. vib.), ionization (Ion) and increase of kinetic energy of the electrons (Kin.). The centre diagram shows the increased ionization due to the Penning effect. (Druyvesteyn and Penning, 1940)

within about 5Å, an electron is extracted from the electrode and the ion is neutralized. The energy liberated is $I - \phi$ where ϕ is the work function for the surface. This energy may cause the secondary emission of an electron if $I - \phi > \phi$, i.e. if $I > 2\phi$. Now for many metals $\phi \approx 3$ V, and so the process is normally possible. However, the efficiency is low: the energy may also be radiated or dissipated in the cathode.

In a Geiger or proportional counter, if a host of argon ions strikes the cathode, some electrons will be emitted, and the counter will recycle. The replacement of argon ions by alcoholic ions is most beneficial because the latter do not provoke secondary emission at the cathode. Rather, the energy liberated on neutralization initiates the dissociation of the poly-atomic molecule. This is one aspect of "quenching action".

A second aspect of quenching is the use of polyatomic molecules to absorb photons emitted in a discharge. The danger always exists in a pure monatomic gas, that electromagnetic radiation will liberate unwanted photoelectrons from the cathode, or elsewhere in the gas itself. Large molecules are often added to the active gas because they are excellent absorbers of the delinquent ultra-violet photons. In this case, the available energy promotes the decomposition of the molecule. For example, the ultra-violet photons emitted by excited argon atoms lie in the range 1070 Å to 790 Å, and they are therefore quenched in the presence of methane, which shows continuous absorption from 1450 Å downwards. There is advantage to be gained in using the largest suitable molecules: for they will decompose many times before being reduced to diatomic molecules. Thus the quenching action of alcohol lasts longer than that of methane. The absorption wavelengths of various gases are presented in Table 1.3.6.

TABLE 1.3.6

Absorption spectra of some polyatomic gases (from Brown, 1959)

Gas	Absorption wavelength (Å)	Type of spectra	Gas	Absorption wavelength (Å)	Type of spectra
Br_2	1500	Continuous	C_2H_2	2400–2090	Bands
Cl_2	1500	Continuous	CCl_4	4600–2300	Bands
CO_2	1360–600	Bands	CH_4	< 1450	Continuous
CS_2	3800–1200	Bands	$CHCl_3$	< 2200	Continuous
H_2S	1600–1190	Bands	CH_3I	3600–2110	Continuous
NO_2	5700–2200	Bands		2100–1215	Diffuse
N_2O	3000–1760	Continuous	C_2H_5Br	2850–1900	Continuous
	1520–1056	Bands		< 1700	Diffuse
	< 1000	Continuous	C_2H_5Cl	< 1700	Continuous
SO_2	3800–1529	Bands	C_2H_5N	< 2500	Continuous
H_2O	1240–983	Bands	C_2H_5OH	1633–1602	Bands
NH_4	1620–1450	Bands		1518	Diffuse
	< 1200	Continuous		< 700	Continuous
			CH_3COCH_3	3300–2940	Bands

1.4 MOTION OF IONS IN THE ELECTRIC FIELD

1.4.1 Collisions, energy distributions

In a gas, electrons, ions and atoms continually collide. If the mean free path is l, and if, say, there are n electrons/sec/cm^2 passing through the gas, then $n\,dx/l$ collisions will be made in distance dx. If these are lost

$$dn = -\frac{n\,dx}{l}$$

and the number remaining is

$$n = n_0\,e^{-x/l} = n_0\,e^{-\mu x},$$

where $\mu = 1/l = NQ$ and $n = n_0$ at $t = 0$. N is the number of atoms/cm^3, and Q the effective cross-section (cm^2). Calculated cross-sections for electron collisions in gases (culled from Meek and Craggs, 1953) are given in Table 1.4.1. The values are for a pressure of 1 mm Hg and 0°C—when the value for N is $3\cdot 56 \times 10^{16}$:

TABLE 1.4.1

Gas	Cross-section (Q cm$^2 \times 10^{16}$)
He	2·83
Ne	4·14
Ar	6·46
Kr	7·54
Xe	9·16
O$_2$	6·9
N$_2$	7·8
H$_2$	3·7
H$_2$O	5·8

In a gas at a certain temperature T, classical kinetic theory tells us that the atoms, ions and electrons will attain an equilibrium that is described by the Maxwellian distribution law:

$$f(v) = \frac{4}{\sqrt{\pi}}\left(\frac{m}{2kT}\right)^{3/2} v^2\,e^{-mv^2/2kT}$$

or

$$f(\varepsilon) = \frac{2}{\sqrt{\pi}} (kT)^{-3/2} (\sqrt{\varepsilon}) e^{-\varepsilon/kT},$$

where $\varepsilon = \tfrac{1}{2} mv^2$, v is the velocity of electron, molecule, etc., and the mean energy $= (3/2) kT$ (Fig. 1.4.1.1).

If electrons are moving in a gas under the influence of an electric field, they may still attain a steady distribution. A necessary condition is that the potential difference per mean free path of the electron El is small in comparison with the electron energy ε. For elastic collisions between electrons and atoms in an electric field, Druyvesteyn and Penning (1940) have shown that a new electron distribution results on account of the relatively small fractional energy loss $2m_e/M$ at each collision with an atom. The distribution is

$$f(\varepsilon) = C(\sqrt{\varepsilon}) e^{-0.55(\varepsilon^2/\bar{\varepsilon}^2)},$$

where $\bar{\varepsilon}$ is the mean electron energy. It is shown in Fig. 1.4.1.1 where it is seen that the number of fast electrons is smaller than for the Maxwell distribution.

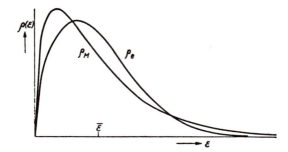

Fig. 1.4.1.1 The Maxwellian electron energy distribution (ρ_m) and its modification by Druyvesteyn. (Druyvesteyn and Penning, 1940)

In practice, the electrons and atoms do not always behave like the smooth spheres of kinetic theory. For example, collisions resulting in excitation and ionization cause the electron to lose energy in large amounts (especially in monatomic gases). The effect is to enhance the number at low energies (Fig. 1.4.1.2). In addition there is the Ramsauer-Townsend effect. This concerns the variation of the collision probability (P) for electrons

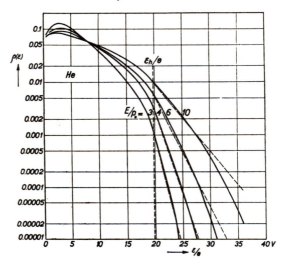

Fig. 1.4.1.2 Electron energy distribution in helium for various fields, after excitation and ionization processes have been accounted for. (Druyvesteyn and Penning, 1940)

in the low-energy region; diffraction as a result of the wave nature of electrons causes an anisotropy in the elastic scattering. As a consequence, the high-energy tail of the electron distribution is depressed.

At this point, it is useful to define P as the number of collisions made in travelling 1 cm through a gas at 1 Torr, 0°C. We have

$$P = \frac{1}{l_0} = NQ$$

or

$$Q = 0{\cdot}281 \times 10^{-16}\, P(\text{cm}^2).$$

We remember that Q can involve the sum of terms for the different collision processes: elastic scattering, ionization, excitation, attachment, etc. We note too that for a gas at p Torr, the mean free path is l_0/p.

1.4.2 Kinetic theory, drift and diffusion

In this section we shall summarize some results from the kinetic theory of ideal gases. They are useful in attempting to describe

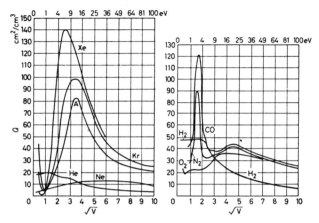

Fig. 1.4.1.3 Measured total cross-sections for scattering of electrons with energy V electron volts, in various gases. (Francis, 1960)

the behaviour of swarms of ions and electrons in discharges, but at the same time they must be treated with some reserve.

In an ideal gas, it is assumed that the particles are elastic spheres. The laws of conservation of momentum and energy apply in collisions, and it is possible, by integrating over all angles, to show that the mean fractional loss of energy at each collision is

$$\Delta = \frac{2mM}{(m + M)^2},$$

where m and M are the masses. In the case of ion–atom or atom–atom impacts ($m \approx M$) we have $\Delta = \frac{1}{2}$, and hence an energetic ion soon loses its energy in a gas.

On the other hand, electrons lose energy slowly in collisions with atoms or ions because

$$\Delta = \frac{2m}{M} \quad (M \gg m).$$

This fact allows electrons in gases to attain high energies under the influence of electric fields, and to avoid being rapidly thermalized. Furthermore, it can be shown that the scattering of electrons off atoms is isotropic, rather than forward as is the case for ions.

When electrons exist in an electric field (E) they experience a force Ee which gives them an acceleration Ee/m in the field direction. This motion will be superimposed on their natural thermal velocities (v). A swarm of electrons of mean energy $\tfrac{1}{2}m\bar{v}^2$ will move towards the anode with a drift velocity u (N.B. $u \ll \bar{v}$). In travelling a distance x in the field direction, the actual path is $x\bar{v}/u$; and the energy lost in collision will be equal to the energy imparted by the field. Hence

$$\frac{2m}{M} \cdot \tfrac{1}{2}m\bar{v}^2 \cdot \frac{x\bar{v}}{u} \cdot \frac{1}{l} = Eex,$$

where l is the mean free path. To take the simplest case, an ion moves in the field direction an average distance $x_1 = ut_1$ between collisions, under an acceleration Ee/m. But the time between collisions $t_1 = l/\bar{v}$, hence

$$x_1 = u\frac{l}{\bar{v}} = \frac{1}{2}\left(\frac{Ee}{m}\right)\frac{l^2}{\bar{v}^2}$$

and

$$u = \frac{el}{2m\bar{v}} E.$$

This is only a crude approximation; more properly the electron drift velocity is

$$u = \frac{2el}{3m\bar{v}} \cdot E.$$

And in the case of ions with a non-isotropic scattering distribution, the appropriate expression becomes

$$u = \frac{el}{M\bar{v}} \cdot E.$$

The factor multiplying E is known as the mobility μ:

$$u = \mu E.$$

(N.B. sometimes K is used for mobility; and sometimes w for drift velocity.)

In practice, this simplified account of drift velocities appears valid at low values of E/p. As E/p is increased (~ 20 V cm^{-1} Torr^{-1}), the mobility ceases to be constant (Loeb, 1955), and at still higher values the drift velocities are better described by the equation

$$u = \text{constant} \times \left(\frac{E}{p}\right)^{\frac{1}{2}}.$$

In any gas, atoms, ions or electrons will diffuse from positions of high to low concentration. The process is described by Fick's

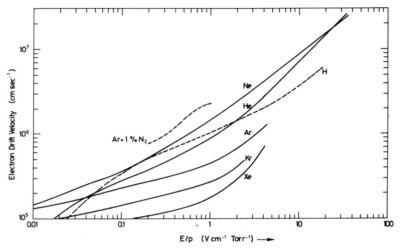

Fig. 1.4.2.1 Electron drift velocities in hydrogen and the noble gases for various values of E/p. [Bowe (1960), Tan *et al.* (1970), Pack and Phelps (1961), Phelps *et al.* (1960) and Brown (1959)]

law of diffusion, which states that the flow of particles, expressed as a flux density, is proportional to the concentration gradient:

$$J = -D\nabla n.$$

from which it is possible to deduce the time dependent diffusion equation

$$\frac{dn}{dt} = D\nabla^2 n.$$

For a volume of particles undergoing irreversible outward diffusion, the concentration at time t will be given by

$$n = n_0 \exp(-t/\tau),$$

KINETIC THEORY, DRIFT AND DIFFUSION 51

where n_0 is the original concentration, and τ an effective time constant.

Kinetic theory shows that the diffusion coefficient D may be expressed as

$$D = \left(\frac{v^2}{3\nu}\right)_{average} = \frac{1}{3}l\bar{v} = \frac{1}{3}\frac{l_0}{p}\bar{v}$$

where ν is the collision frequency.

An implication of this formula is that the electrons in a gas will diffuse far more than the ions on account of their larger mean free paths (l) and velocities (v). The ratio of their diffusion coefficients may be 10^3, and even more in the presence of electric fields.

Einstein's relation follows (roughly) from the above expressions:

$$\frac{u}{D} = \frac{Ee/m\nu}{v^2/3\nu}$$

whence

$$\frac{\mu}{D} = \frac{3e}{mv^2} = \frac{e}{kT}$$

where k is Boltzmann's constant, and T the absolute temperature. And so

$$\frac{D}{\mu} = \frac{2}{3e}\left(\frac{3kT}{2}\right) = \frac{2}{3}\bar{\varepsilon}.$$

(Warning: the above equations express relationships; they are a rough guide; the actual numerical factors are dependent on the distributions, etc.—see Brown (1959), Hasted (1964)).

Townsend showed that the concentration of particles that have diffused from a point source to a radial distance r, at time t is

$$n = \left(\frac{n_0}{4\pi Dt}\right)^{3/2} e^{-r^2/4Dt},$$

where n_0 is the concentration at the source at time $t = 0$ (i.e. Gaussian). Further, the mean square distance of diffusing particles from a source, given by

$$\overline{r^2} = \frac{\int_0^\infty r^2 n 4\pi r^2 \, dr}{\int_0^\infty n 4\pi r^2 \, dr}$$

may be shown to reduce to

$$\overline{r^2} = 6Dt.$$

This represents three-dimensional, i.e. spherical, diffusion. The two-dimensional case, cylindrical diffusion, is given by

$$\overline{r^2} = 4Dt,$$

and the one-dimensional case by

$$\overline{x^2} = 2Dt.$$

The radius of cylindrical diffusion is of particular importance because it is the lateral diffusion of electrons that determines the breadth of many discharges. It may dictate the cross-sectional area of the current, and will be indicated in photographs.

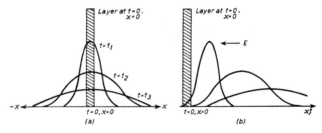

Fig. 1.4.2.2 Diffusion of an electron layer (*a*) in field-free space and (*b*) in an electric field. (After Llewellyn-Jones, 1957)

Figure 1.4.2.2 shows the diffusion of electrons from a layer as a function of time. The effect of an applied electric field is also shown, and the drift is seen to be superimposed on the diffusional motion.

In a thermal plasma containing electrons and positive ions, the electrons will diffuse out at a faster rate. However, if the densities be high enough, the attractive force on the electrons exerted by the positive ions left behind slows down the electron diffusion. The result is that the diffusion flow of both ions and electrons is equal—a phenomenon named "ambipolar diffusion". Under ambipolar conditions, the time dependent equation becomes

$$\frac{dn}{dt} = D_a \nabla^2 n$$

which applies alike to ions and electrons. If the concentrations of ions and electrons are equal, the coefficient is given by

$$D_a = \frac{D_- \mu_+ + D_+ \mu_-}{\mu_+ + \mu_-}.$$

A phenomenon that sometimes affects drift velocities is known as "clustering". Some experiments have yielded mobility values that indicated very high ionic masses. The accepted interpretation is that the ions travelled with a loosely held cluster of neutral (but polarized) atoms surrounding them.

In an electric field, the actual velocities (i.e. the agitation energies) of the ions, and especially the electrons, are increased. Effectively, their temperature rises. For example, in helium with $E/p = 2$ V/cm Torr the mean electron energy rises to about 4 eV which is a hundred times that of the gas atoms (Llewellyn-Jones, 1957). The ratio of electron to gas temperatures as a function of electric field in argon and neon is shown in Fig. 1.4.2.3.

An expression for the average energy of an electron in an electric field can be obtained (Tan et al., 1970):

$$\bar{\varepsilon} = \tfrac{1}{2} m \bar{v}^2 = \frac{Eel}{\sqrt{(3\Delta)}},$$

i.e.

$$\bar{\varepsilon} = \frac{E}{p} \frac{el_0}{\sqrt{(6m/M)}},$$

from which it follows that

$$\bar{v} = \frac{(el_0)^{\frac{1}{2}}}{(1\cdot 5 \, m^3/M)^{\frac{1}{4}}} \left(\frac{E}{p}\right)^{\frac{1}{2}}$$

and

$$D_- = \frac{e^{\frac{1}{2}}}{3} \left(\frac{l_0}{p}\right)^{3/2} \frac{1}{(1\cdot 5 \, m^3/M)^{\frac{1}{4}}} E^{\frac{1}{2}},$$

i.e.

$$D_- \propto E^{\frac{1}{2}} \text{ at constant pressure.}$$

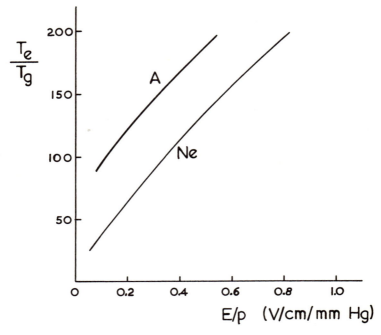

Fig. 1.4.2.3 Ratio of electron to gas temperatures in argon and neon. (Healey and Reed, 1941)

The drift and diffusion of ions may be modified in the presence of a magnetic field. The tendency of the electrons to spiral around the magnetic lines reduces the diffusion perpendicular to B by an amount

$$D_r = D/(1 + (\omega\tau)^2)$$

where the electron cyclotron frequency $\omega = eB/m$; and the time between collisions $\tau = 1/\bar{v}$. Here $\bar{v}$ is the mean random electron velocity (von Engel, 1964). In typical chamber conditions ($B = 1$ Wb. m^{-2}, $p = 760$ Torr) the correction turns out to be small because $\omega\tau \ll 1$, indicating that at these pressures the many randomizing collisions per revolution destroy the effect (see also Chapter 10).

1.4.3 Growth of an avalanche, Townsend theory

In the electric field of a spark chamber, an electron will accelerate towards the anode. It is likely to gain sufficient energy to cause ionization when it collides with a gas atom in its path. An additional electron is thus liberated which, after acceleration, will be able to ionize too. This, and further multiplication, results in an avalanche quickly building up. Travelling rapidly ($\sim 10^7$ cm s^{-1}), the electrons move towards the anode and away from the lethargic positive ions—which move oppositely at approximately 10^5 cm s^{-1}. The process continues until there are something of the order of 10^8 electrons in the head of the avalanche, at which point, it is thought, space charge effects become dominant and the streamer mechanism sets in.

The avalanche process may be described with Townsend's first ionization coefficient (α). It is defined as the number of electrons produced in the path of a single electron in travelling 1 cm in the field direction. The number of electrons produced by n electrons in a distance dx will be

$$dn = n\alpha \, dx,$$

and hence the number produced by one electron in length x is

$$n = e^{\alpha x}.$$

A crude expression for the coefficient α may be obtained from kinetic theory. For an avalanche to proceed successfully, some electrons must have sufficiently long free paths (x_{ion}) to acquire the energy necessary to ionize a gas atom on collision, i.e.

$$eEx_{\text{ion}} \geq V_{\text{ion}} e,$$

where V_{ion} is the ionization potential. Assuming that if an electron gains this energy it will ionize on collision, the chance that it will is determined by the probability of paths greater than x_{ion}.

The number n of electrons that have free paths $> x_{\text{ion}}$ is given by

$$n = n_0 e^{-x_{\text{ion}}/l} = n_0 e^{-V_{\text{ion}}/lE},$$

where n_0 is the number of free electrons, l the mean free path.

The coefficient α is the number of free paths multiplied by the chance of a free path being $> x_{\text{ion}}$, i.e.

$$\alpha = \frac{1}{l} e^{-V_{\text{ion}}/lE}.$$

But
$$l = \frac{1}{Ap}$$
where p is the pressure, A a constant for a particular gas. Hence
$$\alpha = Ap\, e^{-ApV_{ion}/E}$$
or
$$\frac{\alpha}{p} = A\, e^{-Bp/E},$$
where A and B are constants. Some values for A and B are given in Table 1.4.3, together with the range of E/p for which the expression is valid. More exact expressions for α have been derived, which include consideration of the probability of ionization by electron collision and also the electron distribution (Meek and Craggs, 1953). Another approach is to construct an empirical formula that fits the measurements of the first ionization coefficient. For example, Evans (1969) has found the expression
$$\frac{\alpha}{p} = 6.6 \times 10^{-4} \left\{ \left(\frac{E}{p}\right) - 1.42 \right\}^2 /\text{cm Torr}$$
$$\left(\frac{E}{p} > 1.42\right)$$
to be satisfactory.

TABLE 1.4.3 (Badareu and Popescu, 1965)

Gas	A (cm^{-1} Torr^{-1})	B (V cm^{-1} Torr^{-1})	E/p range of validity (V cm^{-1} Torr^{-1})
He	3	34	20–150
Ne	4	100	100–400
Ar	14	180	100–600
Kr	17	240	100–1000
Xe	26	350	200–800
H_2	5	130	150–600
N_2	12	342	100–600
CO_2	20	466	500–1000
air	15	365	100–800
H_2O	13	290	150–1000

GROWTH OF AN AVALANCHE, TOWNSEND THEORY

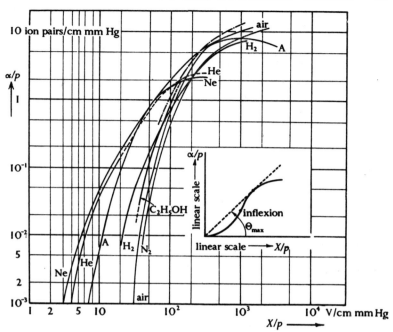

Fig. 1.4.3.1 The Townsend first ionization coefficient α as a function of electric field and pressure for different gases. (N.B. the mean ionizing path $l_{ion} = 1/\alpha$.) (von Engel, 1965)

Some values for α are shown in Fig. 1.4.3.1. Typically, a spark chamber operates in the regime of 10 kV/cm and 760 Torr. Thus the region of E/p of interest is about 13. The figure shows that the number of ions produced in neon is considerably higher than in argon at this field.

An alternative coefficient (η) describing ionizing power is sometimes useful. It is related to α by

$$\eta = \frac{\alpha}{E} \text{ (volt}^{-1})$$

and is the number of ionizations that occur per volt.

Some values for η are drawn in Fig. 1.4.3.2, where the enhanced ionization (Penning effect) can be seen for small additions of argon to neon.

Raether (1964) has taken some beautiful photographs of

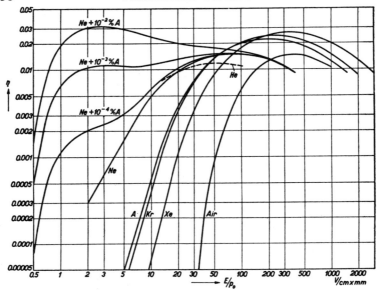

Fig. 1.4.3.2 The ionization coefficient η ($= \alpha/E$) as a function of field and pressure for the noble gases. Here, p_0 refers to the pressure standardized at 0°C. (Druyvesteyn and Penning, 1940)

avalanches (Fig. 1.4.3.3). His apparatus consisted of a pair of plane electrodes 3·6 cm apart, mounted in a cloud chamber containing CO_2 at 150 Torr. A rectangular voltage pulse V, of length

Fig. 1.4.3.3 Cloud-chamber photograph of a single avalanche in CO_2 at 150 Torr (Raether, 1964)

GROWTH OF AN AVALANCHE, TOWNSEND THEORY

250 nsec, was applied to the electrodes after the chamber expansion. The avalanche developed for the duration of the voltage, and at its cessation the free electrons attached themselves to molecules. The positive and negative ions then acted as condensation nuclei for droplets, which were then photographed.

The photograph shows that the avalanche is wedge-shaped, with a rounded head. The length is dictated by the electron drift velocity in the field

$$l = u_- t = \mu_- E t,$$

and the radius by the free electron diffusion

$$r^2 = 4Dt.$$

Recently, Evans (1969) has made a quantitative three-dimensional study of the growth of an avalanche in neon in a field of $E/p = 10$ V/cm Torr. Using the empirical formulae (above), the electron drift velocity

$$u = 2 \cdot 6 \times 10^6 \left\{ \left(\frac{E}{p} \right) - 1 \cdot 5 \right\}^{\frac{1}{2}},$$

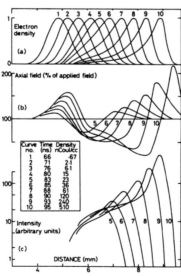

Fig. 1.4.3.4 Variation of electron density, electric field and light output along the axis of a single avalanche during its growth. The units for the top graph are given in the table. The results refer to neon at 690 Torr and $E = 6 \cdot 73$ kV/cm. (Evans, 1969)

and the diffusion coefficient
$$D = 9\cdot 86 \times 10^6/p \quad \text{cm}^2\,\text{s}^{-1},$$
he obtained the graphs of Fig. 1.4.3.4 which show the manner in which the electron density, the axial electric field, and the light output vary with time. The growth is traced in ten stages—from 66 nsec to 95 nsec after the initiation with a single pair—during which time the electron number grows to about 6×10^9 electrons.

Important features to observe in the developed avalanche are the very reduced axial field within the avalanche and the sharply enhanced field at the head. The field also increases towards the rear, although this region is 5 mm downfield from the starting-point. Figure 1.4.3.5 depicts the longitudinal cross-section of the avalanche showing contour lines of electron density and also the field strength. Here we may note how the electrons diffuse laterally and how the field is reduced to the side of the avalanche.

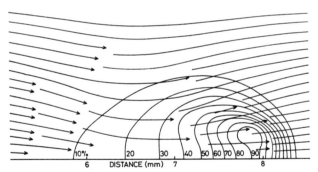

Fig. 1.4.3.5 Longitudinal cross-section of a single avalanche showing contour lines of electron density. The field strength is indicated by the density of lines of force. The diagram refers to the data of Fig. 1.4.3.4 at 90 nsec. (Evans, 1969)

1.4.4. Growth of a streamer

Up until recently, experimental investigations on the development of streamers have been bedevilled by interactions with electrodes. Naturally, in the absence of proper measurements, no agreed quantitative treatment has existed. In the last few years matters have improved. Excellent Russian studies, employing streamer chamber technology, have given reliable data on the space and time characteristics of streamers in neon (Davidenko et al., 1969), and a new model has been proposed by Lozanskii (1969). It is too early to say whether the model will be generally accepted, and indeed whether it will have a general validity in many gases. So, to introduce the subject, we shall first present a simple (and possibly erroneous) picture.

As the avalanche progresses, the space-charge fields of the clouds of electrons and positive ions become important. When the numbers of electrons in the head approaches 10^6, the avalanche begins to slow down owing to the attraction of the positive ions; when it is 10^8 the electrons are much restrained (Raether, 1963).

At 10^8, too, the space-charge field in the middle of the avalanche is so increased practically to negate the applied field, and the neighbouring field is modified as though by a dipole. As a consequence, recombination occurs within the avalanche and ultra-violet photons are emitted isotropically (Fig. 1.4.4.1). It is thought that these photons ionize molecules in the region surrounding the primary avalanche.

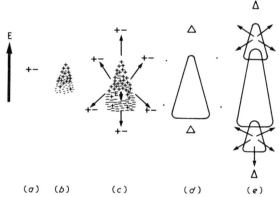

Fig. 1.4.4.1 Stages in the growth of a streamer: (*a*) creation of initial seed electron and positive ion; (*b*) rapid acceleration of the electrons results in the formation of an avalanche; (*c*) growth of the avalanche until the internal field negates the applied field, and recombination causes photons to be emitted which produce more ion pairs; (*d*) the electrons released at head and tail of the initial avalanche grow into avalanches; (*e*) continuation of process with merger of avalanches and growth of streamer.

The seed ion-pairs created in front of the head and behind the tail of the initial avalanche will find themselves in an enhanced electric field, while those to the side will experience a reduced field. As a result, those electrons freed before and aft will accelerate quickly—for α depends on E—to form new avalanches. These will grow until they too become large enough to repeat the process. The old and the new avalanches merge into each other to form a "streamer" whose extremities advance towards the anode and the cathode at a speed of about 10^8cm s^{-1} (in neon under normal spark chamber conditions). On arrival, the electrodes are connected by what is essentially a low resistance conducting plasma of electrons and positive ions. A current, i.e. a spark, will subsequently pass between the electrodes.

A number of points deserve to be made about this picture:

(*a*) The space-charge dipole field reduces the external field to the side of an avalanche, and thus any electrons liberated to the side by photoionization will be accelerated only slowly. For practical purposes, they do not participate in the discharge.

(*b*) We note the photoionization mechanism is not essential to explain streamer growth towards the anode—and indeed Lozanskii and Firsov (1969) discount it. Nevertheless, to continue with our model, we see that the electric field at the head of the initial avalanche is much enlarged by the presence of the electrons just behind, and one would expect it to be greater than the field at the tail (see Fig. 1.4.3.5). This would imply that the secondary avalanches created at the head would develop faster than those behind, i.e. the streamer front advancing towards the anode should be faster than the cathode-directed front. This does happen in some gases, as has been shown by von Tholl's (1963) photographs of discharges in a nitrogen–methane mixture. On the other hand, it does not happen in neon; the early photographs of streamers by Chikovani (1963) showed that the ends of a neon streamer advance more or less equally.

In a fine study of streamer growth in pure neon, Davidenko *et al.* (1969) have used an electronic shutter technique to photograph the development of a single discharge at one nanosecond intervals! (Fig. 1.4.4.2). From their measurements with different fields the graphs of Fig. 1.4.4.3 were obtained. It is seen that at

Fig. 1.4.4.2 The development of a streamer in neon. From the right, the traces show successive stages in the growth of the streamer. The photographs were taken with an optical system including an electronic shutter and an image intensifier. The interval between each exposure was 0·5–1 nsec with an exposure time of about 1·5 nsec. (Davidenko *et al.*, 1969)

a field of 9·6 kV/cm the velocities of both ends of the streamers are the same. At higher fields the cathode-directed streamer is actually faster! Furthermore, it is seen that the velocities increase as the streamer grows. For example, at 16·7 kV cm^{-1}, in growing from 2·5 mm to 6 mm in length, the reverse streamer velocity rises from 0·8 to 2·0 $\times$ 10^8 cm s^{-1}.

(*c*) Streamers exhibit interesting shapes. The photograph (Fig. 1.4.4.4) of the start of a streamer in a 70% neon/30% helium mixture shows a narrow neck from which the streamer appears to be advancing symmetrically in a conical fashion

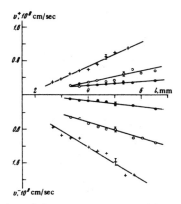

Fig. 1.4.4.3 Velocity of the streamer extremities. The anode-directed (v^+) and cathode-directed (v^-) velocities are expressed as a function of streamer length. The data (dots, circles and crosses) were for field strengths of 9·6, 12·1 and 16·7 kV/cm in neon, respectively. The notable result is that $v^- > v^+$. (Davidenko *et al.*, 1969)

Fig. 1.4.4.4 A single streamer in neon-helium (70/30) showing the symmetrical shape and the neck near the point of origin. (Mishra, 1969)

towards the electrodes. In pure neon the neck does not exist at 10·7 kV cm^{-1}, but with an alcohol impurity a dark gap appears (see Fig. 1.4.4.5). Davidenko *et al.* (1969) have found that a gap decreases from 2·5 mm to 0·9 mm when the field is increased from 10·7 to 17·8 kV cm^{-1}. They inferred that the dark gap corresponded with the critical size of the avalanche at the moment of transition to a streamer.

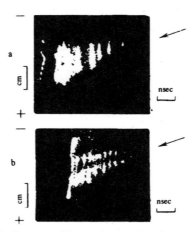

Fig. 1.4.4.5 The influence of impurity on the shape of the streamers. The streamers initiated along the particle track (arrow) are viewed in quick succession from the right. The top discharge was in pure neon, the lower in neon with alcoholic adulteration. (Davidenko *et al.*, 1969)

(d) Raether (1964) has suggested that a criterion for deciding whether an avalanche has developed enough to create a streamer might be

$$\alpha x_c = 20,$$

where x_c is the critical length of the avalanche. This is based on the assumption that the transition to a streamer starts when the internal space-charge field within the avalanche is equal to the applied field. Assuming, roughly, that all the electrons [exp (αx)] are distributed in a sphere at the head of the avalanche, and that the field at the centre of the avalanche will be the field at the radius of the sphere,

$$E_r = \frac{e \cdot \exp(\alpha x_c)}{r^2}.$$

Using the relations

$$r^2 = 4Dt = \frac{4Dx^c}{u_-}, \quad D = \frac{l\bar{v}}{3}$$

and

$$eV_{th} = \tfrac{1}{2}m\bar{v}^2,$$

Raether obtained

$$\exp(\alpha x_c) = 6 \times 10^7 \, V_{th} x_c.$$

From his cloud chamber measurements, he took a value of $V_{th} = 1\cdot 5$ V for air and obtained approximately

$$\exp(\alpha x_c) = 10^8 \, x_c.$$

Hence

$$\alpha x_c = 18\cdot 4 + \log_e x_c,$$

which for normal values of x_c reduces to

$$\alpha x_c \approx 20.$$

This criterion may be used to determine whether a spark gap is large enough (i.e. $>x_c$) to allow a streamer discharge to develop. In spite of its limitations, the criterion has proved to be a useful rule of thumb.

Davidenko et al. (1969) have suggested that $\alpha x_c = 20$ is not always true. In particular, if an impurity is present in the neon, the transition from avalanche to streamer occurs for a smaller electron concentration in the head of the avalanche, i.e. $\alpha x_c < 20$. Schmitt et al. (1969) have developed Raether's approach and on the basis of calculations believe that the streamers are likely to commence when the space-charge field has become 0·7 to 0·8 the value of the applied field—corresponding to about 10^7 electrons in the avalanche.

(e) Some question remains whether any potentially ionizing radiation can escape from within the avalanche. The appropriate absorption coefficients must be such as to allow some photons to pass several mm at least.

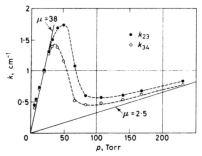

Fig. 1.4.4.6 Two components of the gas-ionizing radiation from a spark discharge in O_2. The absorption coefficient $k(= \mu_{760} \cdot p/760)$ was measured at two distances, 23 and 34 mm ($\equiv k_{22}, k_{34}$). The high-energy component ($\mu_{760} = 2\cdot 5$ cm^{-1}, $\lambda \leqslant 1000A$) may play a role in streamer propagation. (Raether, 1964)

In his book, Raether (1964) discusses the case of oxygen. Taking the results of Przybylski (1958), he argues that of the two observed components of the absorption coefficient at atmospheric pressure (μ) 38 cm^{-1} and 2·5 cm^{-1} (see Fig. 1.4.4.6), the first is rapidly filtered out, but the second may be capable of playing a role in streamer propagation. The value of 0·25–0·5 cm^{-1} obtained for argon (Bainbridge and Prowse, 1956) would support this assertion.

On the other hand, Lozanskii (1969) has pointed to a theoretical coefficient of 300 cm^{-1} in hydrogen for absorption of photons just above the ionization energy, and quoted experimental values of the order of 300 cm^{-1} for nitrogen and oxygen. With these values the ionizing radiation would not emerge from the avalanche. Before the question can finally be decided, more comprehensive and detailed data are required. In the meantime, Lozanskii has made an alternative hypothesis; and the development of streamer chambers has proceeded apace.

(f) The photoionization process requires some consideration. One has to ask how photons from cascade transitions due to recombination within the avalanche can ionize outside. Even if the photons penetrate to the outside they will not have enough energy to ionize atoms like their parent atoms.

If impurities are present in the gas, a feasible sequence might be:

inside the avalanche:

$$He^* \to He + h\nu$$

then photon outside:

$$h\nu + N_2 \to N_2^+ + e.$$

Alternatively outside:

$$h\nu + He \to He^*$$
$$He^* + N_2 \to N_2^+ + He + e.$$

It is known, however, that streamers are created in pure neon. Thus some other mechanism must be proposed. One suggestion put forward by Allkofer (1969) is that the reactions go as,

inside:

$$e + X \to X^{+*} + e + e$$
$$X^{+*} \to X^+ + h\nu$$

outside:

$$h\nu + X \to X^+ + e.$$

It is not easy to appraise this idea. If any cross-sections exist for the production of excited noble ions, they are not easy to find. Hasted (1964) does say, however, that the cross-section rises linearly with the bombarding electron energy. The proposal has the virtue that the photons would not resonantly excite neutral atoms, and hence their escape from the avalanche should be possible. But see the Lozanskii model (section 1.4.5).

1.4.5 The Lozanskii theory of streamers

In a series of papers, E. D. Lozanskii has devised a theory of streamers based on the concept of the streamer as an expanding conducting plasma (Lozanskii and Firsov, 1969). Two growth mechanisms are postulated: the expansion towards the anode is due to the very high field with the leading electrons ionizing furiously, and at the cathode-end photoionization liberates electrons in a high field and these proceed to multiply and move into the plasma. If the quasineutral plasma is truly conducting, its expansion will cause a continuing rise in the field between it and the electrodes.

The model assumes that the equipotential plasma takes the form of two elongated ellipsoids of revolution with sharp boundaries and major axes along the field E_0 (Fig. 1.4.5.1). The field at the extremities (which have a radius R) will be

$$E_a = E_0 a/R \ln \left(\frac{2}{e}\sqrt{\frac{a}{R}}\right), \quad e = 2\cdot 178$$

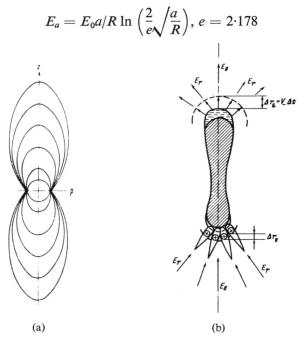

Fig. 1.4.5.1 (*a*) Streamer geometry assumed in Lozanski and Firsov (1969) model. (*b*) Development of oppositely-directed streamers. (Rudenko and Smetanin, 1972)

where a is half the streamer length ($a \gg R$). Calculations yield: the advance velocity,

$$|\dot{a}| = \mu E_0 a/R \ln\left(\frac{2}{e}\sqrt{\frac{a}{R}}\right),$$

where μ is the mobility; the streamer length,

$$a = \exp\left[(\mu E_0 t/R + C)^{1/2}\right]$$

where C is a constant; and the streamer thickness,

$$\rho_{max} = 0.83\sqrt{(aR)}.$$

In the case of a finite conductivity, a field E' will exist within the streamer, and the field at the extremity becomes

$$E_a = (E_0 - \bar{E}')a/R \ln\left(\frac{2}{e}\sqrt{\frac{a}{R}}\right) + E'.$$

Lozanskii and Firsov also calculated the fraction (θ) of the total energy released in a streamer ($E^2/8\pi$) that goes into ionization. In the absence of space charge, the fraction is $\alpha U_i/E$, where U_i is the ionization potential; and in a streamer (with varying fields), θ is obtained with

$$\theta = \frac{8\pi U_i}{E^2 \mu E_0} \int_0^E \frac{\alpha(E)\mu E\, dE}{4\pi} = \frac{2U_i \alpha(E)}{B + 2E}$$

where $\alpha = A \exp(-B/E)$. And using $\theta E^2/8\pi = nU_i e_1$ (n and e_1 are the electron concentration and charge), the field within the streamer can be shown to be

$$E' = 2U_i/\theta a$$

A description of the photoionization mechanism in streamer growth has also been developed by Lozanskii (1969). Although the treatment has not yet withstood the test of time, the fact that it has received experimental support in the case of neon (Davidenko *et al.*, 1969) recommends its inclusion here.

He suggests that the ionization found outside the avalanche, and indeed much of that found inside, is due to the formation of molecular ions (see section 1.3.4) in the collision of excited and neutral atoms:

$$X^* + X \rightarrow X_2^+ + e.$$

The cross-section for this "quenching" reaction at atmospheric pressure is large ($\sim 10^{-15}$ cm^2) and the ionization occurs in times of $\sim 10^{-10}$ sec.

In his analysis, Lozanskii takes into account the broadening of the excitation levels by atomic collisions. This means that instead of the imprisonment, or slow diffusion, of resonance radiation, some photons are able to penetrate large distances (~ 1 cm).

In the following, we shall give an outline of Lozanskii's treatment.

The probability that a photon emitted at the origin is absorbed in a distance r per unit solid angle is

$$w(r) = \frac{1}{4\pi r^2} \int_0^\infty \exp\left[-\mu(\omega)r\right]\mu(\omega)P(\omega)\,d\omega,$$

where ω is the frequency of the photon, $\mu(\omega)$ the absorption coefficient and $P(\omega)$ the shape of the line. Put

$$\mu(\omega) = \frac{\mu_0(\Gamma^2/4)}{(\omega - \omega_0)^2 + (\Gamma^2/4)}$$

$$P(\omega) = \frac{\Gamma/2\pi}{(\omega - \omega_0)^2 + (\Gamma^2/4)}$$

and let

$$x = \frac{2(\omega - \omega_0)}{\Gamma},$$

where Γ is the collision width of the line and μ_0 the absorption coefficient. At the line centre we have

$$w(r) = \frac{1}{4\pi^2 r^2} \int_{-2\omega_0/\Gamma}^{\infty} \exp\left(-\frac{\mu_0 r}{x^2+1}\right) \cdot \mu_0(x^2+1)^2\,dx.$$

In our case $\mu_0 r \gg 1$, $x^2 \gg 1$; and replacing the lower limit by $-\infty$, we have

$$w(r) = \frac{\mu_0}{4\pi^2 r^2(\mu_0 r)^{3/2}} \int_{-\infty}^{\infty} \exp\left[-\frac{1}{x^2}\right] x^{-4}\,dx$$

$$= \frac{1}{(4\pi)^{3/2} r^3 \sqrt{(\mu_0 r)}}.$$

To obtain the number of excited atoms in unit solid angle at a

THE LOZANSKII THEORY OF STREAMERS

distance r, it is necessary to take into account the quenching of an excited atom in an avalanche before the emission of a photon.

It is part of Lozanskii's thesis (which he later supports) that the number of atoms excited to a quenching level is of the same order of magnitude as the number of electrons in the avalanche. The quenching reaction is assumed to be

$$(X^* + X \to X_2^+ + e).$$

If the time for a quenching collision is T, the probability that quenching does not take place during a time t is then

$$w_1(t) = \int_0^t \frac{1}{\tau} \exp\left(-\frac{t}{\tau}\right) \exp\left(-\frac{t}{T}\right) dt$$

$$= \frac{T}{T+\tau}\left[1 - \exp\left(-\frac{T+\tau}{T\tau}t\right)\right],$$

where τ is the lifetime of the excited state.

Combining, the number of excited atoms per unit solid angle at a distance r is

$$N_r = N^* \frac{1}{(4\pi)^{3/2} r^3 \sqrt{(x_0 r)}} \frac{T}{T+\tau}\left[1 - \exp\left(-\frac{T+\tau}{T\tau}t\right)\right],$$

where N^* is the number of atoms excited to a quenching level. N.B. N_r is the number of secondary electrons produced by virtue of the quenching.

In the case of helium at 300°K, quenching will take place from the $n = 3$ level, since the energy of the $n = 2$ state is about 2 eV less than the minimum in the potential energy curve for He_2^+.

The numerical values substituted by Lozanskii were

$$T = \frac{1}{Nv\sigma_T} \approx 10^{-9} \text{ sec}$$

$$\mu_0 = 2\pi N \frac{\omega_{ab}}{\Gamma} \frac{c^2}{\omega^2} \approx 10^7 \text{ cm}^{-1}$$

$$\tau = \frac{2}{3} \frac{e}{mc^3} \omega^2 f \approx 10^{-8} \text{ sec},$$

where f is the oscillator strength (Heitler, 1954).

The times of interest are of the order of 10^{-7}–10^{-8} sec, and so the exponential may be neglected. Then taking $N^* = 10^7$ one obtains $N_r \approx 10$ at a distance of 1 cm, i.e. finite numbers of electrons are produced at large distances from the primary avalanche.

In his paper, Lozanskii proceeds to develop kinetic equations to deduce the contribution that quenching makes to the ionization in the avalanche itself. After a number of approximations, he obtains a ratio of 1:10 for the relative contributions of electron impact and quenching, respectively. Guardedly, he concludes that the quenching makes a contribution to the formation of ions in an avalanche that is at least comparable with that due to ionization by electron impact.

In general the Russian experimental studies have been held to support Lozanskii's model. For example, Davidenko et al. (1969) found a monotonic rise of streamer velocity with length (Fig. 1.4.4.3); and the very existence of streamers in pure neon supports the photoionization mechanism.

Rudenko and Smetanin (1972) have also studied streamers in neon. In Fig. 1.4.5.2, high-voltage pulses (30–100 kV) with a variable rise time of 2–10 nsec and 100 nsec duration were applied to a 38 mm gap chamber. Photographs of the streamers, as a function of time, were obtained by means of a pulsed

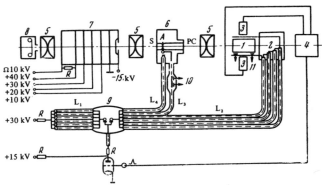

Fig. 1.4.5.2 Apparatus for investigation of streamer growth: (1) spark chamber; (2) cable transformer; (3) scintillation counter; (4) coincidence circuit; (5) Helios 40 objective; (6) electron-optical image converter type ZIM-2; (7) light amplifier; (8) camera; (9) three electrode spark gap; (10 and 11) capacity voltage dividers; PC-photocathode, S–screen, A–anode, L_1–L_4–transmission lines. (Rudenko and Smetanin, 1972)

electron-optical image converter (ZIM-2). The exposures were suitably synchronized with the streamers by deriving the pulses from the same supply, and a synchronized record of the pulses themselves was obtained with the use of capacity-dividers. A timing accuracy of 1·0 nsec and a spatial precision of ±0·2 mm was claimed.

The photographs showed first the growth of the avalanche head towards the anode, then (at 76 nsec in $E = 6·6$ kV/cm the sudden onset of the cathode streamer. At this instant a large current flows in the external circuit; i.e. the oscilloscope registers a sharp drop in the field. An interesting phenomenon occurs at a later stage: a moment arrives when the decline in field strength levels off, and this moment appears to be correlated with an increase in acceleration of the streamer fronts (Fig. 1.4.5.3). (N.B. this work included measurements over a wider range of lengths than the study of Davidenko et al.)

This change in acceleration was interpreted as coinciding with the inception of an instability in the streamer fronts connected with fluctuations in the development of secondary avalanches.

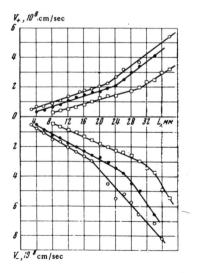

Fig. 1.4.5.3 Streamer velocity as a function of total length, showing the two stages of acceleration in fields: $\bigcirc = 16$; $\bullet = 14·5$; $\square = 10·4$ kV/cm. (Rudenko and Smetanin, 1972)

It is perhaps curious that, given two mechanisms, the instability should commence at the same moment at both ends. It must be noted that Wagner (1966) also observed two stages in the growth of the cathode-directed streamer in nitrogen. This was interpreted as being due to photoionization at the cathode—an explanation that is unlikely to apply in the case of the isolated streamers in the Russian work.

Figure 1.4.5.3 also shows that the cathode-directed streamer fronts advance faster than the anode fronts—which is generally true in fields above 9 kV/cm. Below this value the anode front is faster. The distinct velocities were taken as proof of distinct mechanisms at work.

This section has described the most comprehensive analysis of the streamer so far reported, but it is likely that more remains to be said.

1.4.6 Breakdown; the growth of a spark

A spark corresponds to a current flowing in a gas from one electrode to another. A fine luminous filament, which is normally visible, manifests this conducting plasma of electrons, ions and excited atoms.

In this section we shall describe two classes of breakdown. The first, and slower of the two, relies on the feeding of normal avalanches with secondary electrons ejected from the cathode. The second invokes the streamer mechanism. Before discussing the first, we must remember that the transfer of charge in a spark is dependent on what transpires at the electrodes—unlike avalanches and streamers which occur in electric fields independently of the electrodes. Two secondary processes and their coefficients must be distinguished.

(a) The positive ions on striking the cathode are likely to eject secondary electrons. The coefficient γ is defined as the number of electrons liberated from the cathode per single ionizing collision in the gas (Llewellyn-Jones, 1957).

(b) Photons emanating from the discharge may also release electrons through the photoelectric effect. This is characterized by the coefficient δ such that the number of photoelectrons emitted from the cathode per ionizing collision in a gas is δ/α.

The total electron emission from the cathode due to both effects is then given by

$$\omega/\alpha = \gamma + \delta/\alpha.$$

This is sometimes extended to give the generalized secondary ionization coefficient:

$$\omega/\alpha = \gamma + \delta/\alpha + \varepsilon/\alpha + \phi/\alpha + \beta/\alpha,$$

where ε/α represents cathodic emission of electrons due to excited atoms, and ϕ/α and β/α represent photoionization and ion–atom collisional ionization in the gas (Tan *et al.*, 1970).

As we have noted, a spark between two electrodes does not necessitate the prior development of a streamer. To see that a self-sustaining discharge is possible without it, it is instructive to consider the number (n) of electrons that arrive at the anode, just due to the avalanche mechanism coupled with secondary ionization. Suppose that the cathode is being constantly illuminated with ultra-violet light from an external source, so that n_0 electrons are liberated per second. And, taking the case where electron release by ion bombardment (γ) is predominant, let n_1 be the number so released at the cathode. Then

$$n = (n_0 + n_1) e^{\alpha x}$$

and

$$n_1 = \gamma[n - (n_0 + n_1)]$$

because γ is the average number of electrons liberated by an incident positive ion. Combining, we get

$$n = n_0 \frac{e^{\alpha x}}{1 - \gamma(e^{\alpha x} - 1)}.$$

We notice that we can choose a gap length d such that

$$1 - \gamma(e^{\alpha d} - 1) = 0$$

in which case $n \to \infty$. Now if the illumination ceases ($n_0 = 0$), with a gap d, the nought in the denominator prevents $n \to 0$. Physically, this means that when n_0 electrons leave the cathode, $n_0(e^{\alpha d} - 1)$ ions are created, which on incidence at the cathode liberate $\gamma n_0(e^{\alpha d} - 1)$ [$=n_0$] new electrons (see Llewellyn-Jones, 1957; Meek and Craggs, 1954). This means that the discharge will be self-sustaining after the withdrawal of the source of radiation.

The equation above enables us to determine a breakdown sparking potential. Using the coefficient $\eta = \alpha/E$, the equation becomes

$$\gamma(e^{\eta E d} - 1) = 1$$

i.e.

$$\gamma e^{\eta V} = 1$$

or

$$V = \frac{-\log_e \gamma}{\eta}.$$

The logarithm is negative because γ is much smaller than unity. This expression has been remarkably useful for calculations of static breakdown voltages. If a voltage greater than V is applied to the gap, then α is increased and the current will increase with time—to be limited eventually by the power supply. An important aspect of this breakdown mechanism is the role of the slowly moving positive ions that exist in the gap for significant times. Their space charge has the effect of increasing α, and thus accelerating the breakdown. The time for breakdown by this mechanism is relatively long (>10 μsec) because of the cyclic nature of the process: electrons release ions, which at the cathode release electrons, which release ions, etc., and can amount to hundreds of cycles.

The second breakdown process is via the streamer mechanism. It is very rapid, as it only requires the growth of one avalanche to critical proportions (10^8 electrons) and the two ends of the streamer advance quickly (10^8 cm sec^{-1}) towards the electrodes. On arrival, the conducting plasma then connects the anode and cathode. Presumably, secondary electron emission occurs at the cathode as a result of bombardment by ions and photons. These may provide a feasible explanation for the growth of the spark current to a value in the region of 100 μA. It is not certain, however, what precisely happens at the cathode that enables the current to continue to grow to several amperes in about 10 nanoseconds.

It is well known that at high pressures sparks are contained within fine filaments. From diffusion considerations, it is possible to calculate the area of impact of positive ions on a cathode. For example, with 32 kV across a 1 cm gap in air, the positive ions strike an area of 10^{-2} cm^2 (Llewellyn-Jones, 1957). It is clear that high currents are going to involve intense local heating and evaporation. It is possible that densely ionized plasma standing at the surface of the cathode will reduce the effective work function, but so far no model appears to have been proposed. (The discussion on spark growth is continued in section 5.2.)

1.4.7 The statistics of avalanche growth

Avalanche multiplication has been described by using the coefficient α. On the average, $e^{\alpha x}$ electrons are produced from a single initial electron in a distance x, and the mean length between ionizing collisions is $1/\alpha$. Large fluctuations can occur in the final number of electrons in the avalanche. These are especially dependent on what happens to the primary and early electrons. For example, if it happens that the primary electron goes several lengths $1/\alpha$ before ionizing, the final number of electrons will be considerably reduced. Alternatively, many early ionizing collisions will cause a large final number. The statistics to describe these fluctuations have been developed by Furry (1937), Wijsman (1949) and Legler (1967).

The probability that an electron ionizes an atom to produce a second electron may be expressed as $\alpha\,\mathrm{d}x$ in the range $\mathrm{d}x$. Starting with one electron, we may ask: what is the probability $P(n, x)$ that n particles will result at a distance x? The probabilities have to satisfy the following differential equations:

$$\frac{\mathrm{d}}{\mathrm{d}x} P(1, x) = -P(1, x)$$

$$\frac{\mathrm{d}}{\mathrm{d}x} P(2, x) = -2P(2, x) + P(1, x)$$

....

$$\frac{\mathrm{d}}{\mathrm{d}x} P(n, x) = -nP(n, x) + (n-1)P(n-1, x)$$

....

with the boundary conditions

$$P(1, 0) = 1; \quad P(n, 0) = 0, \quad n > 1.$$

By successive integration and application of boundary conditions Furry obtained

$$P(1, x) = e^{-\alpha x}$$
$$P(2, x) = e^{-\alpha x}(1 - e^{-\alpha x})$$
$$\dots\dots\dots\dots\dots\dots\dots\dots\dots\dots\dots\dots$$
$$P(n, x) = e^{-\alpha x}(1 - e^{-\alpha x})^{n-1}.$$

But $e^{\alpha x}$ represents the mean value of n. Hence

$$P(n, x) = (1/\bar{n})(1 - 1/\bar{n})^{n-1}$$

Or, approximately, if $\bar{n} \gg 1$, we can write

$$P(n) = (1/\bar{n}) \exp(-n/\bar{n}).$$

The formula shows that the probability of an avalanche producing n electrons declines exponentially with n; and also that the fluctuations of n about $\bar{n}$ are large. The top figure (1.4.7.1) shows that under suitable conditions, experimental distributions have agreed with the theory.

The above treatment was based on the assumption of a constant probability of ionization during the avalanche development. Now it must be remembered that after each collision, the electron must travel a distance V_{ion}/E before ionization is again possible. If V_{ion}/E is much smaller than the mean distance between ionizing collisions $(1/\alpha)$, then a constant ionization probability makes some sense—for the velocities are able to achieve equilibrium in non-ionizing collisions. This condition holds for low values of E/p (because the ionization efficiency α/E is small).

However, for high values of E/p (i.e. $V_{ion}/E \sim 1/\alpha$), the assumption of a constant α is very doubtful. It is likely that space-charge effects will occur and that α will be a function of both n and x. Experiments in methylal have shown that the distribution changes from an exponential to a curve with a maximum near $\bar{n}$, as E/p is raised from 70 to 426 $V \cdot cm^{-1} \cdot Torr^{-1}$ (see Fig. 1.4.7.1). So far no satisfactory expression for $\alpha(n, x)$ appears to have been proposed (Legler, 1967). This discussion

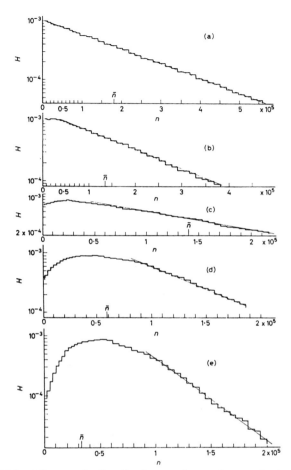

Fig. 1.4.7.1 Measured distributions of numbers of electrons in avalanches. The top diagram shows that the exponential law is obeyed for $k = E/\alpha\, V_{\text{ion}} \gg 1$. The data were obtained in methylal under the following conditions:

Graph	E/p (V/cm Torr)	pd (Torr cm)	d (cm)	k
a	70	53	0·5	26
b	76·5	38·3	0·5	22·6
c	105	13	0·15	10·5
d	186·5	3·4	0·2	5·3
e	426	1·1	0·1	4·1

(Schlumbohm, 1958; Raether, 1964)

1.4.8 Charge transport in noble liquids and solids

The noble gases Ne, Ar, Kr and Xe form simple liquids and solids because the principal binding forces between the atoms are the weak and spherically symmetric van der Waals forces. In both liquid and solid, the local order is determined by the close packing of identical spheres; in the solid, the face-centred cubic configuration is stable. Of special interest for particle detection is the fact that electrons may remain unattached and free to move in the liquid under the influence of electric fields.

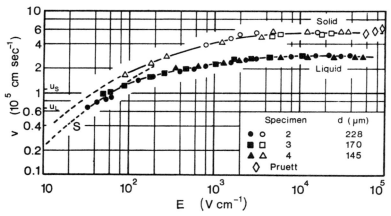

Fig. 1.4.8.1 Field dependence of the electron drift velocity in solid Xe (157°K) and liquid Xe (163°K). (From Miller et al., 1968)

As a typical example, the electron drift in liquid xenon as a function of field is shown in Fig. 1.4.8.1 (Miller et al., 1968). Three regions are discernible: the low field in which the linear relation, $u = \mu E$, holds; the middle where the variation of u is given by $E^{\frac{1}{2}}$; and the high field region in which the drift velocity saturates. The first two regions are reasonably described by the Schockley theory in which the energy distribution of the hot electrons is Maxwellian, with an effective temperature T_e, and where it is assumed that the rate of energy gain by the electron system from the applied field is equal to the rate of loss by acoustic scattering. The saturation velocity has not been properly explained, although Cohen and Lekner (1967) have shown that a mean free path that is dependent on structure becomes important at high fields.

Studies on positive ion mobilities at low electric fields (Davis et al., 1962) have indicated that the charge-transporting ions are Ar_2^+, Kr_2^+ and Xe_2^+ (i.e. not Ar^+, etc.). The diffusion coefficient of an ion is related to the mobility by the Einstein relation, $u_i = D_i q/kT$; and on the other hand, the self-diffusion coefficient of the neutral molecule, $D_0 = A \exp(-B/T)$ where A and B are pressure-dependent parameters. It is found that $D_0/D_i \sim 5$; and the suggested explanation is that the polarization interaction makes a significant contribution to the friction.

TABLE 1.4.8

Low field mobilities for noble liquids and solids from Miller et al. (1968) and Davis et al. (1962). The numbers, especially in the last column, are examples only; they are very dependent on temperature and pressure.

	Electron mobility ($cm^2\ V^{-1}\ s^{-1}$)	Velocity of sound ($10^5\ cm\ s^{-1}$)	$\left(\dfrac{\text{Sat. elec. vel.}}{\text{vel. of sound}}\right)$	Positive ion mobility ($10^4\ cm^2\ V^{-1}\ s^{-1}$)
Ar (solid)	1000	1·38	10	
Ar (liquid)	475	0·85	9·4	6·5
Kr (solid)	3700	1·1	8·0	
Kr (liquid)	1800	0·7	5·4	7·2
Xe (solid)	4500	1·1	5·0	
Xe (liquid)	2200	0·65	4·4	3·0

CHAPTER 2

Historical Review

AT the conception of this monograph, it was intended that the second chapter should be an historical review of the early growth of the family of particle detectors that depend for their action on electron multiplication processes in gases. The story goes back to 1908 when Rutherford and Geiger announced their discovery that in an arrangement of a wire anode mounted coaxially in a cylindrical cathode "the current through the gas due to the entrance of the α particle into the detecting vessel was magnified . . .". This fertile ancestor generated a large family whose more important branches form the subject of this book.

The family has grown in pace with the needs of nuclear and particle physics. Premature births often withered through a lack of demand only to be resurrected when the time was ripe. This interrelation between experimental technique and the state of knowledge is a fascinating aspect of contemporary physics.

It might be imagined that modern history is straightforward—if the writer has lived through the events, how can he misinterpret them? And with all the published work in front of him how can he fail to give a correct account? Of course, such a belief is naïve to a degree. Should present-day history, therefore, be attempted, or should a decent interval of time be allowed to elapse to gain perspective?

The question of accuracy is one that must trouble the scientific historian. This is not true in all fields: the political historian writing, say, the biography of a cabinet minister will inevitably inject his own views; his argument will be contentious, and the polemics that follow publication will provide a further illumination. The debates on television, the dissection in literary journals, the ripostes by offended parties are the very means of extracting historical truth.

Science, on the other hand, calls for a more serious approach. With only poorly developed facilities for subsequent discussion, erratic conclusions based on inadequate research may never be challenged. A grave risk exists for permanently misjudging contributions and injuring reputations. For example, without detailed interviews it is often difficult to identify the originator of an idea: sometimes a student gets the credit for his supervisor's proposals; sometimes the student is exploited by his supervisor.

And anyway, how do creative ideas emerge? Is it important that they might spring from a casual conversation over a cup of coffee, or from a remark in a seminar? What credit are the secondary contributors entitled to?

Any historical treatment of detection chambers should attempt to name the inventors and proceed to assess the significance of each contribution to the mainstream of particle physics. Often this is possible, but occasionally we are presented with the Newton–Leibniz tangle: is it more important to be first with an invention or to invent later but publish in a fashion that excites other physicists to exploit it? And how does one compare the contribution of a lone physicist in a small institution with those of a well-endowed team at a synchrotron laboratory?

After pondering on these questions, the writer's view is that the history of this branch of science is too complex a matter for superficial treatment in one chapter. A whole book is called for —and one based on detailed historical investigation, and interrogation of the practitioners. Hence in what follows, only a few milestones are mentioned, and only those that pointed the way to the successful chambers described in this monograph.

The early history of the spark chamber has been reviewed by Roberts (1961). In this work Keuffel (1949) is identified as the man who first observed that the discharge between parallel plates caused by the passage of a particle is physically located along the path of the particle. Roberts commented that this observation went unnoticed and had to be repeated independently three more times before it penetrated the general consciousness as a useful way of observing particle tracks.

Bella and Franzinetti (1953) published the first photographs of the spark discharge and estimated the precision to be within one cubic millimetre. In Hamburg, Henning (1955) with Bagge introduced the techniques of using several parallel-plate counters; of enhancing the spark with a triggered condenser discharge; and of taking stereo photographs. In his recent

(1969) essay, Allkofer—writing all the more authoritatively for having participated in the group—mentions that Henning achieved a precision of 0·2 mm after having experimented with various mixtures of argon and alcohol. This group was one of the first to do an investigation with spark chambers—on the multiple scattering of cosmic ray muons.

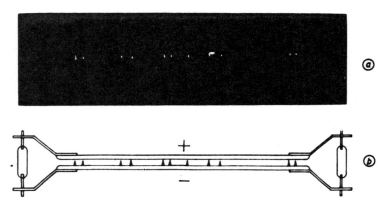

Fig. 2.0.0.1 Localized sparks in an early spark counter. (Bella and Franzinetti, 1953)

At Harwell, Cranshaw and de Beer (1957) introduced the method widely employed today of applying the high voltage to the chamber in the form of a pulse immediately after the passage of a selected particle. By this means the bane of spurious pulsing may be eliminated. A small clearing field of about a hundred volts was constantly applied to sweep away unwanted ions. It appeared that the clearing times could be less than a microsecond which would allow the use of the chamber in high-intensity beams from accelerators. Large multiplate assemblies were constructed but, strangely, only air at atmospheric pressure was used as the operating gas.

At this stage, if many simultaneous particles traversed the chamber only one track would cause a spark. It was Fukui and Miyamoto (1959) who showed that simultaneous tracks could be recorded if noble gases were used and if the rise time of the high-voltage pulse was sufficiently short.

Furthermore, a consequence of using noble gases was that in a reasonable gap—say 2 cm—the sparks actually followed the

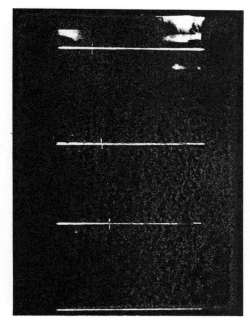

Fig. 2.0.0.2 Sparks indicating a particle trajectory in the original pulsed air-filled six-gap spark chamber. (Cranshaw and de Beer, 1957)

Fig. 2.0.0.3 Photograph of a track in a chamber containing neon plus argon at atmospheric pressure, with a pulsed field of 13 kV applied to the 2 cm gaps. (Fukui and Miyamoto, 1959)

line of the ionized trail, rather than the electric field (Fig. 2.0.0.3). This phenomenon led to the development of the wide-gap chamber in Russia by Alikhanian et al. (1963), in which clear spark trails of 20 cm or more could be obtained.

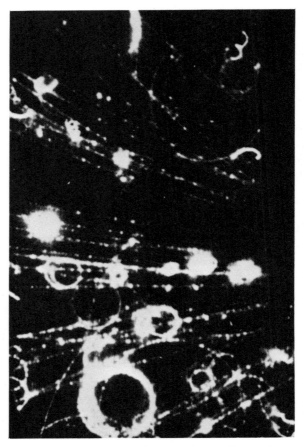

Fig. 2.0.0.4 Early photograph of a cosmic ray shower in a streamer chamber with a magnetic field present. (Chikovani et al., 1965)

Considerable thought was devoted to the mechanism of spark growth in noble gases. It was realized that in a high field the specific primary ions initiated avalanches which subsequently created sparks. The intermediate mechanism however, the

HISTORICAL REVIEW

streamer, was the subject of much speculation. Nevertheless, it was recognized that by using high-voltage pulses of only short duration, it was conceivable that the discharge could be prematurely arrested—i.e. at the streamer stage before it became a spark. It was Chikovani *et al.* (1963) and Dolgoshein *et al.* (1964) who invented this approach and who devised the first streamer chambers. The visible streamers were localized at the avalanche sites, and it turned out that their most effective means of recording was by photography through transparent electrodes. The beautiful photographs in Fig. 2.0.0.4 were obtained in this way.

From these beginnings the modern instruments of today have evolved. This is a suitable point to conclude Chapter 2 because much of the remaining story constitutes this book. However, no historical review, even the briefest, can omit mention of Prof. G. Charpak. For nearly two decades his contributions have been extraordinarily inventive and in recent years he and his group at CERN have been responsible for both the proportional chamber and the drift chamber. Indubitably, Professor Charpak has been pre-eminent in this field.

CHAPTER 3

Narrow-gap Spark Chambers

3.1 AN ELEMENTARY SPARK CHAMBER

SPARK chambers are not the prerogative solely of synchroton laboratories. No mammoth budget is required to finance them, because they may be simple and cheap to build. They make ideal undergraduate experimental projects for they are satisfying, not only cerebrally, but aesthetically too. The vivid sight and dramatic sound of sparks in a chamber responding to a cosmic ray shower is on a par with son et lumiere at Notre-Dame. And the practitioner has the reward that comes to the artist on creating a beautiful happening. It is largely for this reason that a description of an elementary chamber is presented in this section; but it may also be instructive to those encountering spark chambers for the first time.

The basic elements are shown in Fig. 3.1.1.1. The passage of a particle through the chamber is recognized by two scintillation counters which rapidly initiate a process that culminates in the application of a high-voltage pulse to alternate plates of the chamber. A row of bright red (in neon) sparks will indicate the particle trajectory.

The electrodes in Fig. 3.1.1.1 may be fashioned from $\frac{1}{4}$ in. aluminium plates of area 30×30 cm². To prevent spurious

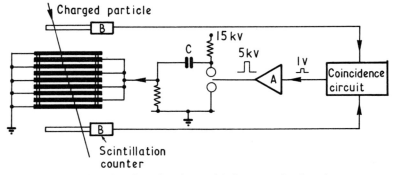

Fig. 3.1.1.1 Sketch of a simple multiplate spark-chamber system.

discharging within the chamber, the edges of the electrodes should protrude outside the sensitive gas volume. This may be achieved by spacing the electrodes uniformly with $\frac{1}{4}$ in. perspex walls that are milled to give a uniform gap dimension of 8–10 mm. At the corners, the walls will have to be joined with perspex cement, for they will contain the gas volume. Indeed, it is essential that the system be gas tight as small traces of air inside the chamber will ruin its performance. This leads to the main practical difficulty: there appears to be no good metal–perspex cement. An alternative method may be used for sealing the tops and bottoms of the walls to the electrodes. One possibility is to use homemade flat rubber gaskets. The necessary pressure may be applied with horizontal wooden joists lying above and below the chamber. If the joists are connected with $\frac{1}{2}$ in. nylon threaded tie-rods, an optimum pressure may be applied to ensure sealing. It is sensible to smear the rubber gaskets with vacuum grease. (N.B. this is not to say that cementing with epoxy resin is impossible: Meyer and Terwilliger (1961) have reported success.)

The surfaces of the electrodes should be flat and smooth and it is sensible to round off the edges. Gas will have to be admitted and rather than have inlet and outlet ports for each gap, a small (1 mm) hole may be drilled in each electrode. To ensure good circulation the holes in alternate electrodes should be drilled at opposite ends of a diagonal. This chamber is unlikely to be vacuum tight, and the best method of introducing the active gas is by flowing it slowly but continuously through the chamber. This may seem wasteful, but it prevents a common experience, which is the poisoning of the gas by vapours emanating from the walls and plates of the chamber.

A satisfactory gas to use is the 70% neon, 30% helium mixture supplied by British Oxygen Company at about 3p a litre. At the start of operations, it is economical to exclude the (electronegative) air from the chamber by first flushing thoroughly with argon. A gas system must be devised so that when the air has been driven out, the neon–helium from a cylinder may be switched in at approximately atmospheric pressure. It is good practice to maintain a slight excess pressure so that if there is a small leak the gas flow will tend to be outwards. This excess may be maintained by causing the gas to bubble through a chosen height (say, 6 cm) of silicon oil—both before the inlet and after the outlet. Practice will indicate the best rate of gas flow, although flow meters are available.

Then there is the triggering. Cosmic rays passing through the chamber may be detected with pieces of plastic scintillator (say, $20 \times 20 \times 1$ cm^3) placed above and below. The light pulses from the scintillator, following the passage of a particle, may be transferred to the photocathodes of photomultipliers via (1 cm or $\frac{1}{2}$ in.) perspex light guides shaped as equilateral triangles. Both the plastic scintillators (say, NE 102A) and the cement may be bought from Nuclear Enterprises, Edinburgh. In a simple project it is sensible to use the slower venetian blind type of photomultiplier (say, from EMI, Hayes, Middlesex) rather than the faster focused dynode variety, for the longer pulses are easier to handle. All the plastic and light guide should be shrouded with shiny aluminium foil and then given some accident-proof and light-proof outer cover.

The fact that one is creating fast voltage pulses of 15 kV or so, means that special shielding and earthing arrangements have to be made if transistor circuits are employed in the detection of coincidences and in the amplification. Otherwise the transistors blow up. An alternative is to restrict oneself to valves throughout, for they are much more robust and satisfactory in the present context. Following the passage of a charged particle through both scintillators, simultaneous (coincidence) pulses from the photomultipliers may be detected with a single 6BN6 valve. This valve has two independent control grids, and will pass a current if positive pulses occur at both grids, but not otherwise. If thirteen-stage venetian blind multipliers are used, with the recommended manufacturers circuit, the pulses may be taken from the twelfth dynode rather than the anode. In this way one may obtain positive rather than negative output pulses. The simultaneous arrival of two such pulses, fed directly through a capacitor to the coincidence valve, results in a negative pulse being produced at the anode of the valve.

The next stage is the amplifier. It is required to amplify the negative 2 volt pulse up to about 5000 V. It is important to perform the amplification with as little delay as possible; and the output pulse should have a fast rise time. A suitable circuit is shown in Fig. 3.1.1.2.

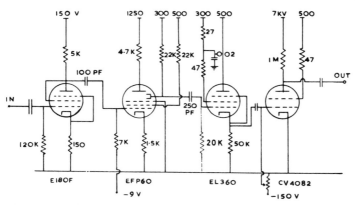

Fig. 3.1.1.2 Trigger amplifier. A two-volt pulse is amplified to 6 kV before application to a spark gap. (Rice-Evans and Mishra, 1969)

The output pulse may be directly applied to a tungsten needle whose tip lies between two spark-gap electrodes made of 1 in. diameter stainless-steel ball-bearings. If the tip lies just above (2 mm) the earthed ball, a small spark will occur whose photons will proceed to breakdown the complete gap (≈ 1–2 cm). The purpose of this breakdown is to short the positive high-voltage plate of the capacitance (C) to earth, thus forcing the charge on the opposite plate to be shared with the alternate plates of the spark chamber. This happens because the resistance prevents its quick escape to earth. One wishes to ensure that the pulse that appears on the spark-chamber plates has a fast rise-time. To achieve this it is necessary to use a rapid-discharge capacitor C—i.e. one with a low inductance [they can be obtained from Wego Co.]. A suitable value is 5000 pF.

The pulse may be applied to the chamber with coaxial cable. A good way to make the connection is via a stub of 1 mm wire that fits tightly into a 1 mm hole drilled in the edge of the particular plate. The earthed sheath of the coaxial cable will have to be stripped back a few inches from the point of connection.

If a single coaxial lead be used, and the alternate plates interconnected by conductors closely situated, it will be found that often only one spark will occur in a ten-gap chamber. This is because spark growth is a statistical affair—one spark will grow before the others, and the entire available charge will pass to earth via the spark channel, before the other sparks have had time to develop. One way to prevent this is to provide each plate with its own coaxial lead of several metres length. They will all be connected together at the capacitor C, but the lengths will introduce sufficient delays to ensure that the voltage is maintained for a long enough period on each plate to allow sparking in all gaps. An alternative method of isolating one gap from another is to connect the plates to the capacitor via individual series resistances. The cable lengths or the resistance values are best determined by trial and error.

To reduce the problems of voltage surges around the laboratory, it is wise to connect the earthed plates of the chamber via an earth loop back to the spark gap. (The loop is not shown in the figure.) It is also sensible to isolate all the apparatus, including the power supplies, from the mains with a filter that will not transmit high frequencies.

Typical cosmic ray tracks in this chamber are shown in Fig. 3.1.1.3 (Little *et al.*, 1966).

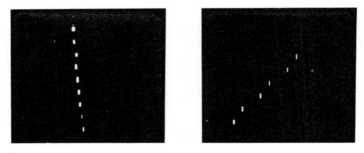

Fig. 3.1.1.3 Cosmic-ray tracks in a spark chamber. (Little *et al.*, 1966)

3.2 RUDIMENTS

3.2.1 Spark formation

There is a criterion to be satisfied if satisfactory sparks are to be obtained in a spark chamber. It concerns the size of the interelectrode gap and the amplification in the gas. A proper spark channel will be created only if streamer action occurs. As was explained in section 1.4.4, this happens when at least 10^8 electrons have been produced in the avalanche head. This number leads to the Raether criterion

$$\alpha x_c > 20.$$

From this we see that for a given α (the first Townsend coefficient), the minimum distance from the anode at which a seed ion-pair may be created and still initiate a spark is $20/\alpha$. The electrode spacing (d) must therefore be greater than $20/\alpha$. For a particular chamber with given d, the applied field must be chosen to fulfil the condition.

In the formation of a spark, the major fraction of the time taken is due to the build up of the avalanche; the subsequent transition to a spark takes a much shorter time. Therefore, if the average drift velocity of the electrons be u_-, using Raether's criterion, the spark formation time will be

$$\tau = 20/\alpha u_-.$$

An alternative estimate has been given by Fischer and Zorn (1961), and although it is open to considerable doubt it is of sufficient interest to include here. They took as their starting-point the theory of Dickey (1952) in which the avalanche is considered to grow up to the size that corresponds to the current passing in the spark. One may consider a spark chamber of capacitance C, in series with a generator V_0 and a resistance R. If at time $t = 0$, a step voltage V_0 appears across the gap (d), then one may write for the voltage across the plates while the avalanche is forming

$$V(t) = V_0 - \frac{1}{C}\left[\sum_{i=1}^{N}\frac{e}{d}(x_{\bar{i}} - x_{\bar{i}}^{+}) - \int_0^t I_{\text{ext}}\, dt\right]$$

where $I_{\text{ext}} = (V_0 - V)/R$ is the current supplied by the external circuit. This current replaces the energy that has been removed from the electrostatic field; the gap voltage will drop when it fails to cope. N is the number of electrons produced, $x_{\bar{i}}^{+}$ is the location of the ith positive ion relative to the cathode measured in the field direction and $x_{\bar{i}}$ is the location of the ith electron. Differentiating this expression, and neglecting the positive ion movement, leads to

$$C\frac{dV}{dt} = -\frac{euN_0}{d}\exp\left(\int_0^t \alpha u\, dt\right) + I_{\text{ext}}$$

whence, if α and u_- are treated as constants,

$$\tau = \frac{1}{\alpha u} \ln\left[\frac{I_{\text{ext}}\,d}{N_0 e u}(1 + RC\alpha u)\right].$$

To test this formula, Fischer and Zorn devised an experiment with the parameters $I_{\text{ext}} \sim 5\,\text{A}$, $R = 20\%$, $C = 30\,\text{pF}$, $d = 0.66\,\text{cm}$. N_0 was taken as one, and α and u were taken from known values at the various fields. Insertion of these values gives

$$\tau = \frac{(29.5 \pm 0.5)}{\alpha u}$$

where the ± 0.5 accommodates the maximum variation in He, Ne, A over the range of experimental voltage.

This estimate clearly differs from the previous estimate. The extent to which the formula is supported may be seen in Fig. 3.2.1.1 where both the measured and calculated times are plotted. Fair agreement is found for argon and helium, but not for neon. Calculating backwards, the time $29.5/\alpha u$ implies an avalanche with 10^{13} electrons; which contradicts the streamer

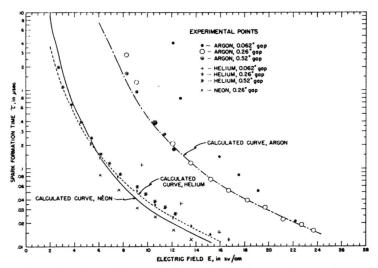

Fig. 3.2.1.1 Comparison of the calculated and observed spark-formation times (τ) as a function of applied field (E), for different gases and gap sizes. (Fischer and Zorn, 1961)

picture of spark growth. However, the actual numerical values for the spark formation times are not drastically affected because of the exponential. The practical importance of the formation times is that they indicate how long the pulse must remain on the chamber plates for the spark to develop fully.

In the Rutherglen (1964) graph (Fig. 3.2.1.2) the two estimates of the critical length ($18·4/\alpha$ and $29·5/\alpha$) are expressed as a function of the applied field. Although the former is more acceptable, one would wish anyway to employ a gap size that allowed for a margin of error and for fluctuations in the deposit of primary ionization.

Rutherglen (1964) has discussed the relevance of spark formation times to variations in gap size. It had been commonly observed in multigap chambers that quite small errors in the spacings resulted in large variations in the relative spark intensities. If two gaps have different spacings, the field (and hence α) will be less in the larger gap. The consequence will be that the time for spark development will be shorter in the smaller gap. If the same high-voltage plate is serving both gaps, a spark will occur in the small gap, but not the larger.

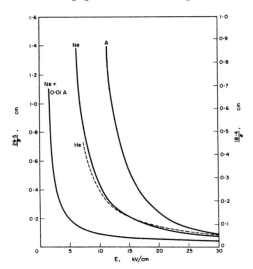

Fig. 3.2.1.2 Critical lengths $18·4/\alpha$ and $29·5/\alpha$ for the formation of avalanches of 10^8 and 10^{13} ion pairs as a function of applied field at atmospheric pressure. (Rutherglen, 1964)

Rutherglen extended Fischer and Zorn's treatment and obtained the expression for the ratio of the numbers of electrons in the two gaps

$$\frac{N_2}{N_1} = \exp\left(30(\beta + \tfrac{1}{2})\frac{\Delta d}{d}\right)$$

where Δd is the variation in spacing and $\beta = (E/\alpha)(\mathrm{d}\alpha/\mathrm{d}E)$ which can be obtained from graphs of α vs E. The number 30 is the 29·5 of Fischer and Zorn. As an example, take 20 kV/cm in neon for which $\beta = 1\cdot 5$. Then if $\Delta d/d$ is 1/60 one obtains $N_2/N_1 = e$, a sizeable difference in intensity for a 1·6% variation in spacing. Even more strikingly: a 15% difference in spacing results in an intensity ratio of e^9, so it is not surprising that no sparks are seen in this instance.

3.2.2. Techniques for construction

Narrow-gap sampling chambers have been built in a wide variety of fashions. The essential requirements are: (*a*) a gas-tight sensitive volume; (*b*) a uniform interelectrode gap; (*c*) good insulation between the electrodes; (*d*) no spurious breakdown due to field emission or conducting paths; (*e*) a means for recording the information.

If photography is to be used, at least two walls of each gap have to be transparent. Perspex is the easiest to machine and it also has excellent insulation properties. Precautions have to be taken to ensure that spurious discharging does not occur at the edges of the electrodes. One solution to this problem is to bring the electrodes out into the air by using perspex spacers as the chamber walls. This is very effective because the breakdown potential is so much higher in air, than in the chamber gas. The method suffers from the disadvantage that the high voltages appear on the exterior, although a second perspex box can be built to contain the whole chamber.

A second solution is the shelf-type of chamber in which the electrode edges lie in grooves that have been milled in the perspex walls. Burleson *et al.* (1963) have employed this method with success. Thin electrodes were required with dimensions 70×30 cm^2. They were prepared by chemically etching 1 mm aluminium plates down to a thickness of 0·075 mm, except for a border of 1·5 cm around the edges. Thicknesses down to 0·04 mm were prepared in the same way. To prevent bowing of the electrodes due to gas pressure differentials, the ends of the chambers were enclosed with thin sheets of mylar, sealed with a silastic adhesive.

If the electrodes are not too thin their edges may safely exist in the sensitive gas if they are carefully rounded. This method was employed in the CERN neutrino spark chamber (Faissner *et al.*, 1963). A modular principle was adopted: each unit of two gaps was constituted by two outer earthed electrodes and one central high-voltage plate. The outer electrodes (160 × 100 × 0·5 cm^3) and some metallic spacers were glued together with araldite to form a rigid frame. The central electrode was suspended with small perspex insulators. The front face was closed by an optically polished perspex window. All parts were sealed off with cold-setting silicon rubber compound. The whole spectrometer consisted of 80 aluminium units and 100 brass units, the total weight amounting to 25 tons!

The majority of spark chambers for research have been bespoke items; they have been designed with particular experimental requirements in mind. As a consequence, they have assumed many physical forms—e.g. electrodes have been made consisting of thick steel, brass and aluminium plates, very thin aluminium, conducting glass (e.g. glass painted with SnO_2), printed circuit boards and wire grids. Fischer and Zorn (1961) have described a technique for mounting thin aluminium foils tautly on a frame. They start by clamping the foil between two concentric rings, and then placing it on a flat circular plate of smaller diameter and with a rounded edge for stretching. The matt surface of the aluminium was placed upwards for sealing with epoxy to a perspex frame. It is useful to seal at a temperature about 10°C below normal room temperature, to exploit the fact that the thermal coefficient of expansion for perspex is about three times that for aluminium, and thus the subsequent warming up will enhance the stretching.

It is not essential that a spark chamber should have plane electrodes. A spherical or cylindrical geometry would perform well as long as the gaps were uniform. Beall *et al.* (1963) have described a ten-gap cylindrical chamber whose electrodes were composed of 0·01 in. rolled aluminium. The chamber was 18 in.

long and the electrodes, which varied in diameter from 10 to 20 in., were supported in circular grooves machined in polished perspex end-plates.

3.2.3 Chamber gases

It was the attention paid by Fukui and Miyamoto to the gas fillings that led to the general application of spark chambers in high-energy physics. By employing a noble gas as the main amplifying agent, the chambers acquired a reliable performance and a capacity to handle simultaneous particles. The distinguishing characteristic of the nobility is the absence of electron affinity. Free electrons do not attach themselves to form heavy ions and so the multiplication can proceed unimpeded. Faster spark formation times result and the consequent reduction in timing fluctuations allows multiple sparks to develop simultaneously.

In choosing the optimum composition for a spark-chamber gas, several characteristics must be considered—in particular: the Townsend ionization coefficient (α), the drift velocity, the electron affinity, the photo absorption, the specific ionization, the diffusion coefficient, the emission spectrum, and indeed the cost. The main gas is restricted to the nobility—i.e. to helium, neon, argon, krypton and xenon. At an E/p value of about 10 V cm^{-1} Torr^{-1}, Fig. 1.4.3.1 shows that neon and helium have the highest Townsend coefficient α. The implication, which can be seen in Fig. 3.2.1.2, is that for a particular gap size a smaller electric field may be applied. Of the two, neon in general is to be preferred because its total specific ionization is higher (Table 1.2.2) and thus the chance of a satisfactory spark in a small gap is much improved.

In practice, for reasons of cost, a mixture of neon and helium is commonly used. The proportions tend to vary with the country: 90/10, 70/30, 65/35 all appear satisfactory. It is generally thought that these commercial grade gases will contain a fraction of argon too. This is most helpful, for the Penning effect (see section 1.3.5) will occur, and neon atoms excited to metastable states will contribute to the overall ionization.

Although the random presence of impurity gases will have a

most deleterious effect on the chamber performance, the controlled introduction of certain additional gases and vapours may be advantageous. One instance is the application of small quantities of electronegative gases; e.g. a little freon will not attach the primary electrons sufficiently quickly to render impossible the avalanche production; but over a longer time scale it will help to clear the chamber of free electrons.

It may be seen in Fig. 1.4.2.1 that the addition of 0·1 % nitrogen to argon increases the electron drift velocity by a factor of about five. The excited rotational and vibrational states of the N_2 molecule are much lower than the excited states of noble gases. As a consequence, electrons suffer inelastic collisions before they achieve high energies and so the electron distribution is concentrated at low energies (temperatures). However, at these lower energies (~ 1 eV) the Ramsauer–Townsend effect comes into play: the scattering of electrons by noble gas atoms is affected by the wave nature of the electrons and a longer mean free path (i.e. a greater drift velocity) results. This effect promotes the removal of electrons from the chamber by the permanent application of a d.c. voltage.

The effect of small additions of N_2 and O_2 to argon has been studied by Hohne and Schneider (1963), and of air to neon by Burnham and Thompson (1964). Also the discussion of spark-chamber performance (see section 3.5) will include many references to gaseous effects.

The presence of organic molecules such as alcohol may be beneficial because they are efficient absorbers of photons. They will fulfil a quenching action by preventing photoionization at sites in the chamber a long way from the original avalanche. They will, of course, be electronegative and attach electrons, but even this has been turned to advantage by Kotenko et al. (1967). They showed that the attachment reduced the diffusion of electrons away from the true path of the particle during the delay before the application of the field; and yet they could be detached again for avalanche production if the applied voltage was sufficiently high.

The importance of control of gas purity has been emphasized. One way to prevent the poisoning of the active gas by desorption from the walls is to flush the chambers continuously. However, the cost of the gas has some bearing on the procedure (e.g. in London, the standard 70/30, Ne/He mixture costs 3p a litre). Well-endowed institutions can afford to exhaust it into the atmosphere—even though an ambitious experiment may consume 5000 guineas worth of gas in a year. On the other hand, short experiments of a day or so may be performed with chambers that can be evacuated and then filled. Economies in large systems can be made by continuously purifying and recycling the gas. For example, McLaughlin and Schafer (1969) have designed a system including molecular sieves that will remove all nitrogen, oxygen, water, alcohol and other vapours, and transmit only neon, helium and hydrogen. A simpler purification procedure, that would freeze out the organic molecules, would be to pass the gas through a cooling trap of $-70°C$ before entry to the chamber. It must be said, however, that the commercial gases are good enough for normal spark-chamber operation. Loeb's (1955) recommendations for gas purification are valuable.

3.3 WIRE PLANES

3.3.1 Principle

A "wire plane" is a spark-chamber electrode that is comprised of many parallel wires. Typically, wire planes may have an area of several square metres and consist of a thousand or so 0·1 mm diameter stainless-steel wires each separated by 1 mm. One end of each wire would be connected to a bus-bar through which the spark current is taken.

With the development of fast computer techniques for data handling and on-line analysis, spark chambers incorporating wire planes have become increasingly popular. Their advantage lies in the fact that when a spark occurs, the current is localized in one or two wires, and a knowledge of these wires determines one of the coordinates of the spark. A single gap created by two wire planes mounted orthogonally, will give both the x and y coordinates. Alternatively, two gaps, each formed from a plate and a wire plane, will do the same.

3.3.2 Multiple sparks

When many simultaneous sparks occur, ambiguities may arise. In Fig. 3.3.2.1(*b*), spark currents in wires, X_1, X_2, Y_1 and Y_2 do

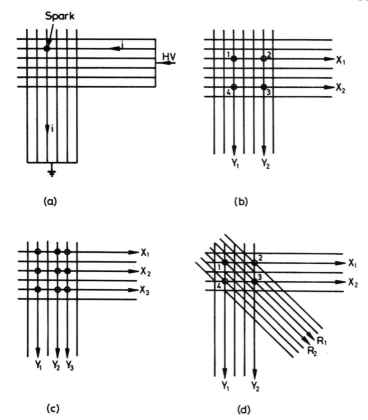

Fig. 3.3.2.1 Spark coordinates obtained with crossed wire planes. Ambiguities that may arise and the use of a third plane.

not indicate the two sparks uniquely—they tell only that the sparks may have been at 1,3 or at 2,4. Figure 3.3.2.1(c) shows how the uncertainties increase when more than two sparks occur. The situation is not improved by using many gaps with their wires in the x, y directions; rather, to resolve the difficulties extra planes must be mounted at angles other than 90°. In Fig. 3.3.2.1(d) currents in wires X_1, X_2, Y_1, Y_2, R_1 and R_2 indicate the sparks occurred at 1,3 and not at 2,4.

Large numbers (say, n) of simultaneous sparks require wire planes at $(n + 1)$ angles for complete determination. However, the ambiguities with three planes can be small, if there be large numbers of wires in each plane. Figure 3.3.2.2 from Miyamoto (1964) and Fischer (1967) shows the calculated ambiguities for up to ten spark tracks.

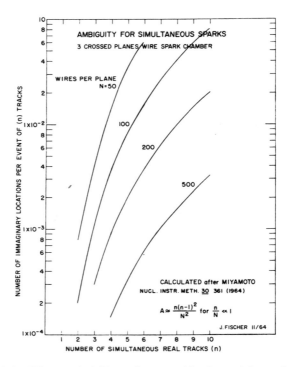

Fig. 3.3.2.2 The probability of an ambiguity arising when three wire planes are used, as a function of the number of sparks. (Miyamoto, 1964; Fischer, 1967)

Although a third grid mounted at 45° would give the least uncertainty, this is sometimes inconvenient since the corners of the frames would not overlap, and also, the sensitive area would be diminished. For some applications where closely lying sparks are not expected, much smaller angles may be used. Schubelin (1968) used an angle of only 5° which was satisfactory for resolving sparks at distances greater than 12 mm from each other—the wire spacing in his case being 1 mm.

3.3.3 Attainable precision

At first glance, one might expect that for a grid of 1 mm wire spacing the possible accuracy of measurement would be 1 mm (i.e. ± 0.5 mm). However, this may be improved upon by allowing the spark current to be of such a magnitude that more than one wire may participate in the discharge. Galster *et al.* (1967) have pointed out that the best procedure is to adjust the high-voltage pulse, and its length, on the chamber so that the average (over a large number of single events) number of wires ($\bar{N}$) that fire is 1·5 (or half-integer). In this case, one pulsing wire indicates that the traversing particle passed very near the wire, while two imply it passed close to the mid-point between two wires. The precision will then be $\pm s/4$ where s is the wire spacing. (N.B. if $\bar{N} = 2$, the accuracy is only $\pm s/2$!) Calculations showed that an average $\bar{N} = 1.5$ could be obtained at high efficiency by employing 5·3 kV pulses with a decay time of 220 ns.

Under reasonable conditions, assuming that the read-out electronics is capable of handling the information; a measurement precision of ± 0.4 mm is obtainable (Zanella, 1968). In fact, Perez-Mendez (1966) has reported an accuracy of 0·3 mm using magnetostriction, and Galster *et al.* (1969) has quoted 0·2 mm with ferrite-core read-out.

3.3.4 Construction

The construction of wire planes is something of an art. For large chambers great care has to be taken to wind the wires with a uniform tension that must be strong enough to maintain tautness and yet insufficient to cause snapping after some weeks of operation. The material and size of the wires and their separation is important: the author has used 0·1 mm stainless-steel wires; and at CERN, large chambers have been constructed with 0·06 mm phosphor-bronze (attempts to use 0·04 mm proved unsatisfactory). It was the experience of the CERN group (Schubelin, 1968) that while a 1 mm wire separation worked well, a 2 mm spacing produced erratic sparking—presumably due to field distortion at the wires.

The winding procedure itself may consist of rotating the frame like a coil with the wire feed being weighted with a pulley at constant tension. On the other hand, to construct very large planes, a splendid device has been constructed at CERN in which the araldite frame lies on a table and a carriage shuttles to and fro, each time advancing 1 mm, while feeding out the wire at fixed tension. At the ends of the traverse the wire is looped over 1 mm sprockets on the sides of the table. Later the wires are araldited to the frame (Alleyn *et al.*, 1968).

The use (Krienen, 1967) of wires that have been "wiggled", i.e. made to have a springiness and hence to retain the tension—has been favoured by some "pioneers", but Schubelin (1968) found that the response of 1 mm bins over 1 mm spaced-wire planes proved non-uniform.

Fischer (1965, 1966) has developed wire planes from etched circuit sheets. These have the advantages: (*a*) that the electrodes can be very thin (hence less likely to scatter particles) and (*b*) that unlike some wires, they do not break. Starting with a large negative of 500 lines (20 per inch, 25 mil wide), Fischer was able to photoprint planes with a substrate of 3–4 mil fibre-glass epoxy and a copper grid either 1·3 or 0·6 mil thick. The same technique has been used to produce planes of 40 lines per inch.

3.4 PULSING THE CHAMBERS

3.4.1 Requirements of a trigger system

The electric field is applied momentarily to a spark chamber with a high-voltage pulse circuit. The requirements of the pulsing system are essentially:

(*a*) that it should apply the pulse as rapidly as possible after the passage of the particle. This is to ensure that the ionization electrons along the track in the chamber have neither been cleared away, nor attached to electronegative atoms, nor allowed to diffuse far from their origin, before the pulse is applied. In addition, a small delay means that the memory time of the chamber may be shortened, and the chance of unwanted sparks reduced.

(*b*) that the pulse has sufficient energy, and lasts for a suitable time, so that satisfactory sparks will be obtained. Where multiple sparking is envisaged, the system must supply an energy to each spark that will be adequate for the recording device.

(c) that the high-voltage pulse rises sufficiently rapidly to ensure spark formation, rather than merely sweeping away the track electrons.

(d) that it should have a short recovery time to allow a rapid operation of the chamber.

(e) that it should be robust and capable of working for hundreds of thousands of pulses without murmur.

The basic action of a trigger circuit is to switch the electrostatic energy stored in a capacitance C_1 (see Fig. 3.4.1.1) into the spark chamber (capacitance C). The capacitance C_1 normally stands charged to a high-voltage V, while both electrodes of the spark chamber are at earth potential. When a pulse arrives from the amplifier, the charge stored in the left-hand plate of C_1 is shorted to earth via the switch. The charge on the right-hand plate then has to share itself between C_1 and the

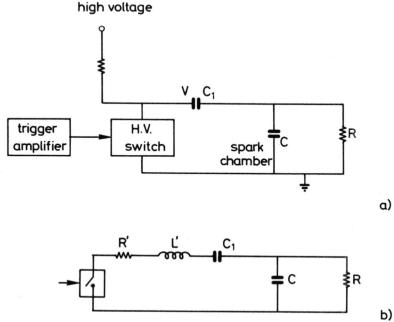

Fig. 3.4.1.1 Basic spark-chamber circuit.

spark chamber C, because the resistance R prevents its rapid escape to earth. The result will be a voltage of amplitude $C_1V/(C + C_1)$ applied to the chamber. In the absence of a spark, the pulse will decay exponentially with a full-time of $R(C_1 + C)$. In the event of a spark, the charge will leak away and the effective pulse length will be shortened.

This simple treatment ignores the inductances and resistances of the system although they can be important when dealing with fast pulses. Cronin (1967) has discussed the example of charging a large chamber capacity of 1000 pF (1 cm gap, 100 pF/sq ft). He estimated that typical values would be several ohms for the series resistance of the thyratron switch, and ~ 0.01 μH/in. for the inductance of the leads. To charge 1000 pF to 10 kV in 10 nsec requires a current of 1000 A. Thus for a 10 in. lead ($L = 0.1$ μH), in applying a 10 kV pulse, the maximum rate of rise of current is $di/dt = V/L = 10^{11}$ A/sec. This current can just rise to 1000 A in 10 nsec. In the equivalent circuit, where L' is the total inductance around the loop, and C' is the sum capacitance, $CC_1/(C + C_1) \approx C$ for $C_1 \gg C$; if R be ignored and if $R' \ll (4L/C')^{\frac{1}{2}}$, then the circuit will resonate with a frequency $1/2\pi(L'C')^{\frac{1}{2}}$ and the rise time of the pulse will be $\sim \pi(L'C')^{\frac{1}{2}}$. This is 30 nsec in the above example, if $R_s \ll 20\Omega$. This last condition would be satisfied by spark gaps whose typical effective resistance is a few ohms (Fischer and Zorn, 1962). In this resonant circuit the voltage rises to twice the value $VC_1/(C + C_1)$.

3.4.2 Production of the high-voltage pulse

The switch that is used to short-circuit the storage capacitor to earth is normally either a thyratron or a three-electrode spark gap. Each of these will pass the necessary large current if they are triggered with a pulse of about 5 kV. It is therefore necessary to amplify the 1 volt signal from the decision-making (coincidence) logic up to this level.

A popular approach has been to use three stages with the valves: E180F, EFP60 and a 6130 (e.g. Cronin, 1967, Fig. 3.4.2.1). The negative 1 volt pulse is first amplified and inverted in the E180F. This is an isolation amplifier and will help protect any transistors in the preceding logic. The pulse is then applied

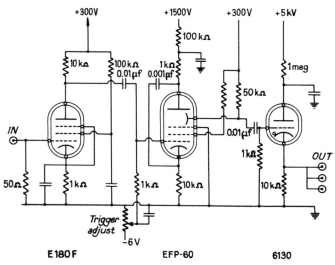

Fig. 3.4.2.1 A typical pulse amplifier for driving spark gaps or thyratrons. (Cronin, 1967)

to an EFP60 which will allow a current pulse of 1 amp to be taken from its dynode. Although this pulse may drive a thyratron switch (say, 5C22) directly, it is sensible to interpose a 6130 thyratron, so that a signal may be taken from the latter's cathode to drive several switches. Alternatively, a hard valve may be interposed as was proposed in the earlier circuit (Fig. 3.1.1.2).

The choice of switch is governed by a number of requirements. It must have a short overall delay time; and it must be able to pass large currents with a very short rise-time. The 5C22 thyratron that has been mentioned has an overall delay of about 100 nsec. The maximum current is about 1000 amp. Cronin (1967) has found that a single 5C22 will drive a 1500 pF chamber with a plate voltage of 15 kV. Bigger systems will require several thyratrons.

On the other hand, spark gaps have some advantages over thyratrons. The breakdown time may be of the order of 10 nsec. This will allow a very fast rise-time, and a short overall delay. By investigating the oscillations in a basic circuit induced by sparks, Fischer and Zorn (1962) observed in their spark gap

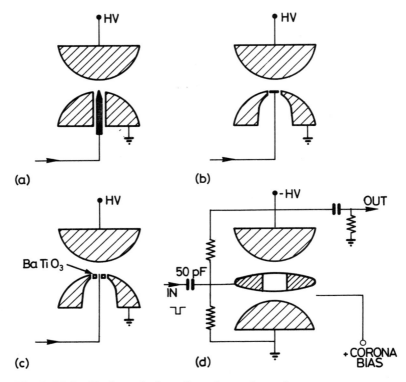

Fig. 3.4.2.2 Various designs for triggered spark gaps.

sparks whose inductance was ~ 3 nH and whose resistance was $\sim 0\cdot 8\ \Omega$. They pointed out that even this small inductance results in a relatively long rise-time for the spark-chamber voltage. For example, with a 500 pF load, a rise-time of about 13 nsec would be expected.

Various designs of spark gap have been used, of which four are shown in Fig. 3.4.2.2. In each case the gap sizes are adjusted so that a constant high voltage may just be maintained across them. In the first design, the negative pulse causes a corona discharge at the needle point. As a result the liberated electrons and ultra-violet photons proceed to break down the main gap. In the second example, the trigger pulse produces breakdown between the central pin and the earthed electrode, and this spark initiates the main breakdown.

The third design (Lavoie et al., 1964; and favoured by the author) relies on the high dielectric constant of barium titanate (~ 2000) to concentrate the electric field just above its surface. Consequently, relatively small trigger pulses (~ 1000 V) will produce the initial breakdown across its surface from the pin tip to earth. This method reduces the necessary prior amplification and may therefore be useful in shortening the delay. The fourth example is the cascade gap in which a third electrode exists centrally at an intermediate voltage. The application of a trigger pulse (say, -5 kV) to the central electrode will swing the potential and overvoltage the lower gap. Breakdown will occur, and emitted photons will pass up through the hole in the middle electrode, to discharge the top gap, and thus turn on the switch. It facilitates rapid breakdown if free electrons are always present in the lower gap. Fischer and Zorn (1962) suggest a constant corona discharge of ~ 2 μA from the etched tip of an adjacent 6 mil tungsten wire.

Spark gaps may be operated in the open air or enclosed and pressurized. Nitrogen is sometimes used. There is an advantage in operating at high pressure because the separation between the main electrodes will be smaller for the same voltage. And a smaller gap means a smaller breakdown time, and hence a shorter overall delay before the application of the pulse to the chamber. A design by Hübbeling (1972) is shown in Fig. 3.4.2.3, in which corona needles are mounted within the central (trigger) electrode.

Although spark gaps have many advantages, experience has shown that their very fast pulse rise-times can cause interference and even the destruction of surrounding electronics. Consequently, at CERN, the switch in general use is a high-power thyratron (English Electric CX1157) which may deliver up to 1000 A with rise-times of the order of 15 to 30 nsec, and which has a useful life exceeding 10^8 pulses. The CERN circuit (Evans and Friend, 1968) is shown in Fig. 3.4.2.4. The thyratron is fired by a stacked avalanche transistor circuit which requires a NIM input signal. Lindsay and Pizer (1968) estimate that the delay between the input trigger and the half height of the output pulse is 60 nsec with a jitter of 1 nsec.

It was stated in the beginning of section 3.4 that a small overall delay means that the memory time of the chamber may

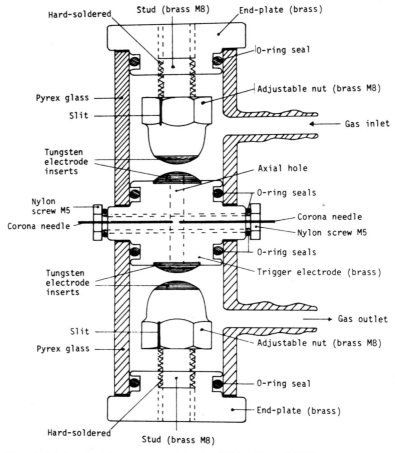

Fig. 3.4.2.3 Spark-gap assembly by Hübbeling (1972)

be shortened. Fischer and Zorn (1962) discussed the lower limit and made the following points:

(a) that in a beam experiment, the triggering counters should be situated upstream so that their signals will travel in the same direction as the particles. (N.B. a common pulse velocity in coaxial cable is 0·9 c, which is equalled by 288 MeV pions and 1937 MeV protons.)

(b) Assuming fast phosphors, advantage can be gained by

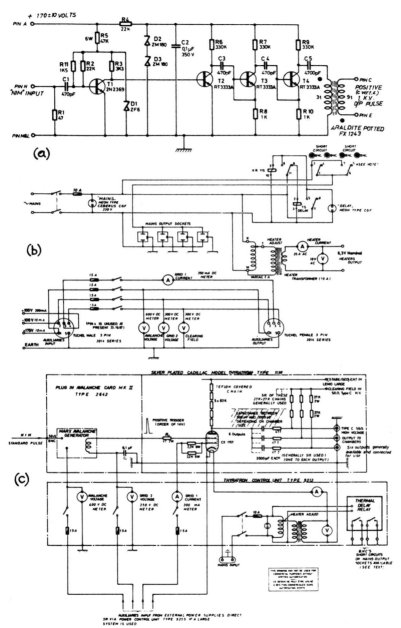

Fig. 3.4.2.4 The standard CERN thyratron HV pulser (up to 20 kV) designed by W. M. Evans and B. Friend (1968)

taking the output from an early dynode in the photomultiplier.
(c) The amplifier might be a high-voltage avalanche transistor (2N2087) delivering 200 V in 3 nsec, followed by the secondary emission dynode valve EFP60 in which a 2 kV anode swing can be achieved with a delay of 5·3 nsec.
(d) By overvolting the chamber (e.g. 25 kV/cm in neon) the spark formation time can be held to about 2·5 nsec.
(e) They concluded that the overall delay in the whole system might be reduced to 33 nsec, as indicated in the table:

Photomultiplier tube	12 nsec
Coincidence circuit	1
Avalanche trigger	3
HV amplifier plus rise-time	6
Spark gap	7
Propagation of HV from gap to centre of spark chamber of 2 ft	2
Spark formation time	2
Total	33 nsec

No one has yet approached this optimistic figure; in practice 200–300 nsec is common.

3.4.3 Application of the pulse to the chamber

A single cable will transfer a high-voltage pulse to a chamber plate. As has been mentioned, when many plates are driven off a single storage capacitor it is sensible to decouple the plates, one from another, so that the first HV plate to spark will not rob the other plates of their charge before they too spark. This may be achieved either with long individual leads to each plate, or alternatively with individual connections via resistors. The high-voltage field propagates with a finite velocity and it is sensible, especially with large chambers, to connect the lead to the plate as symmetrically as possible. A spark chamber behaves like a waveguide and one consequence is that the pulses often reflect from the edges. This may be alleviated with terminating resistors.

In the case of wire chambers, the high-voltage pulse is

normally applied to a heavy copper bus-bar that is connected to all the wires of the plane. Special precautions have to be taken if their advantages are to be exploited. In particular the quality of the spark current must be carefully controlled; for example, when ferrite cores are used for recording it is necessary to prevent overshoot in the current flipping back the core and thereby destroying any information. The current should therefore be limited to a height just sufficient to flip the core (the spark does not have to be visible), and its oscillation damped.

Resistive paste has been a popular means of isolating the wires from each other to prevent robbing. In a design by Galster et al. (1967) the spark current was limited by equipping each wire with its own small storage capacitor, created by separating the HV bus-bar from the wires with a 75 μ Hostophan insulating foil. A variety of pulsing methods have been advocated and a couple of CERN designs are illustrated in Figs. 3.4.3.1, 2.

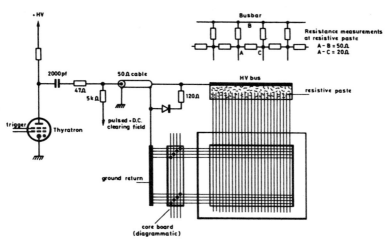

Fig. 3.4.3.1 CERN (Lundby/Baker) wire-chamber system. (See Lindsay and Pizer, 1968)

The voltage rise-time may be reduced by using a distributed system if the spark impedance is small relative to the characteristic impedance (Z_0) of the system. Fischer and Zorn (1962) took this approach and Fig. 3.4.3.3 shows a spark gap situated between a distributed source of stored energy and a matched

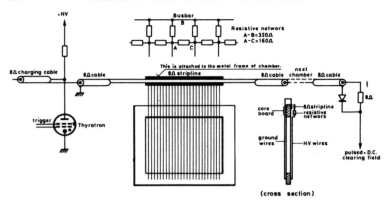

Fig. 3.4.3.2 CERN/Aachen wire-chamber system. (See Lindsay and Pizer, 1968)

distributed load. The load, i.e. the chamber, was terminated by the characteristic impedance— 5 ohm. For a single strip line

$$Z_0 = 377 \, d/(\omega\sqrt{k}),$$

where d and ω are the gap size and chamber width, and k the dielectric constant. Therefore, in the double-gap chamber $Z_0 = 188 \, d/\omega$. The current rises in less than one nanosecond to a maximum, $V_{dc}/2Z_0$, and the maximum pulse voltage is half the d.c. voltage (V_{dc}) and rectangular in shape. The pulse width is twice the transit time of the charging line, and of course must be larger than the spark-formation time.

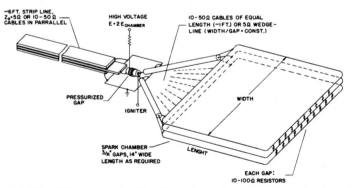

Fig. 3.4.3.3 Elements of a distributed spark-gap and spark-chamber system. (Fischer and Zorn, 1962)

THE CLEARING FIELD

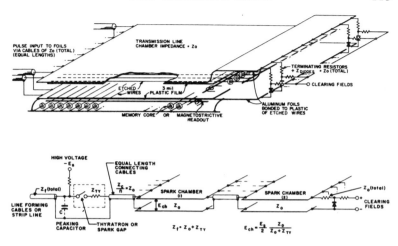

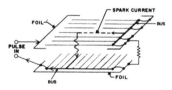

Fig. 3.4.3.4 A transmission line system for pulsing several chambers. The relationship is seen between the pulse input, the pulsed foils, the wire electrode, the read-out system, the connections and terminations. (Fischer and Shibata, 1968)

Figure 3.4.3.4 shows how the transmission line technique may be extended to incorporate the advantages of wire chambers.

3.4.4 The clearing field

A d.c. field in the region of 0–200 volts is normally maintained across spark-chamber gaps to clear the volume of ions from unwanted tracks. By this means the sensitive time can be improved. The rate of clearing will depend on the field and the electron drift velocity. Customarily it is applied in the direction opposite to the applied HV field—to enhance the chamber efficiency (see section 3.5.1). The magnitude of the clearing field is partly determined by the delay between the triggering signal and the application of the HV pulse—the latter must appear before all the desired electrons (corresponding to a selected event) have been drifted from the gap!

After sparking, a larger pulsed clearing field may be employed to clear the gap of plasma electrons. Such a technique improves the recovery time and allows fast cycling of the chamber. To take an example, Faissner et al. (1968) applied a pulsed voltage of 1 kV for 1 millisecond to the chamber immediately after sparking. In Fig. 3.4.3.4 (r.h.s.), to permit the application of this pulse, it was necessary to connect the 8 Ω terminating resistor to earth via seven 10D10 diodes in parallel. And the clearing voltage supply had to be able to withstand the charging current of the pulse-shaping cable.

3.5 CHAMBER PERFORMANCE

3.5.1 Single-track efficiency

The efficiency of a single gap refers to the chance that the passage of a particle will result correctly in a spark. Nowadays, it is possible to achieve very high efficiencies (>95%) in single gaps, so that in a system of many contiguous gaps the overall efficiency is essentially 100%. For example, in eight gaps, if one or two sparks are missing, the trajectory may still be ascertained.

A spark requires the production of a critical avalanche ($\sim 10^8$ electrons) from one or more free electrons. The efficiency of a gap will therefore depend upon the existence of free ion pairs in such a position that the applied field can create the necessary avalanche. It is no good if the field be too low, the rise time too long, or the gap too small; for the electron cloud will merely strike and be absorbed by the anode before the critical size is reached.

The ionization in the wake of a particle depends upon the gap, the particle and its energy. For minimum-ionizing particles, the number of primary ions/cm at atmospheric pressure in helium, neon and argon is about 6, 12 and 30, respectively (Charpak et al., 1965). However, there will be a Poisson fluctuation in this density along the path and so the probability of finding an electron in a space x cm in, say, neon will be $1 - \exp(-12x)$, i.e. 97% for $x = 0.5$ cm, 100% for $x = 1$ cm. If the necessary distance for growth of an avalanche be y, then in a gap of size d, the relevant space (x) is equal to $d - y$ and will be on the side of the gap away from the pulse anode. That is, there is a dead space of depth y adjacent to the anode.

A clearing field is normally applied in a narrow-gap chamber to reduce the memory time. If this d.c. field be in the same direction as the pulse field there is a danger that all the electrons will drift out of the relevant space and the spark will be precluded. On the other hand, if the d.c. field be opposed, electrons will drift into this space—from whence they can initiate a spark.

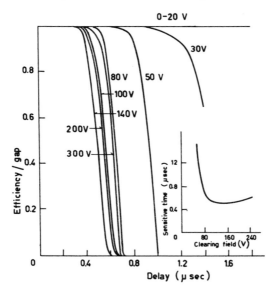

Fig. 3.5.1.1 Single-spark efficiency in a spark chamber as a function of delay time and clearing field. (Cronin and Renniger, 1961)

There is a limit to the time available. For example, Fig. 1.4.2.1 shows that with a clearing field of 200 V/cm the electron drift velocity in neon is 5×10^5 cm/sec; and thus a 1 cm gap will be emptied in 2 μsec. In this case, a safe procedure would be to apply the pulsed field after a 1 μsec delay. Figure 3.5.1.1 shows the results of measurements by Cronin and Renniger (1961) on the efficiency of a single gap filled with pure neon. The relation with clearing-field and pulse delay is seen. It may be noted that the allowed delay is in fact anomalously less than that predicted by conventional values for drift velocity. Another curious feature is that the sensitive time reached a minimum (140 V) and and increased again as the clearing field was raised.

It appears that the composition of the gas has a great influence on the efficiency. Even small impurities will drastically alter the chamber performance. This is due mainly to the attachment of the free electrons by the electronegative adulterations. Many workers (e.g. Cronin, 1960) have observed how superior fresh neon is to stale—and it is for this reason that either flushing, or cleansing (e.g. over charcoal) and recirculating, is common. Burnham and Thompson (1964) have shown how the efficiency drops when small quantities of air are present.

The effects due to small additions of non-noble gases may be put to good use however—but these will be discussed in section 3.5.4 on memory times.

3.5.2 Spatial resolution

It is well known that sparks may be located in narrow-gap chambers with an accuracy of 0·5 mm or better. We shall discuss a number of factors that contribute to the limit of precision.

The electrons created initially along the particle trail will have energies of a few electron volts. They will, therefore, diffuse before the HV pulse is applied. For example, Heyn (1961) found that a delay of 300 nsec in neon resulted in an r.m.s. diffusion distance of 0·8 mm. This diffusion, being statistical, will affect the spark location and will depend upon the ionization density. Rutherglen and Paterson (1961) using three gas fillings (100% He, 75% He + 25% A, 90% Ne + 10% A having primary ionization densities of 5, 12, 14 per cm) found, in the angular range of 15–30°, spreads with standard deviations of $\pm 0\cdot 8$, $\pm 0\cdot 3$ and $\pm 0\cdot 4$ mm, respectively. Broadly, this confirms that greater ionization leads to greater precision (Charpak *et al.*, 1965). The spark width is a function of the energy discharged. Typically it might be 1 mm. The geometrical properties of sparks have been discussed by Burnham *et al.* (1963).

A number of systematic errors are likely to arise. In narrow-gap chambers the sparks are essentially perpendicular to the electrodes, even for angled tracks. However, as was mentioned in the previous section, the sparks are most likely to be initiated by electrons near the pulse cathode. Figure 3.1.1.3 shows the spark displacement that results. The presence of a clearing field introduces another effect: for the initial line of electrons drifts in the direction of the field. This again causes a displacement, but it may be arranged to help cancel the first effect. In any event, where many adjacent gaps with reversed fields are employed, averaging will enable accurate determination. Yet another effect is the drifting sometimes caused by a slow rise on the HV pulse, although this is normally negligible (Astbury *et al.*, 1967).

Other effects might include occasional spurious sparks due to initiation by delta rays, or some photoelectric effect.

Christenson *et al.* (1964) studied spatial accuracy with an optical system that consisted of eight gaps of 0·375 in. each, on which the fiducials were scribed to an accuracy of 0·001 in. The film measurement accuracy was 120 μ in real space. Straight lines were computed from the eight sparks, and the deviations of individual sparks from the lines were estimated for two ranges of track inclination.

The accuracy of track determination, of course, will include the limits imposed by the recording system. A discussion is given in Chapter 4 of the possibilities available with each recording technique. Here we may just observe that the plate chamber with photographic recording has the advantage that it is not digitized; and the possible accuracy has been reported to be about 0·09 mm (Astbury *et al.*, 1967). With wire chambers, a good accuracy would be 0·3 mm, although Fischer (1967) has reported a precision of 0·1 mm with magnetostriction applied to 0·25 wire spacings, with timing by a 100 MHz clock.

3.5.3 Multiple-track efficiency

In many experiments, several simultaneous particle trajectories have to be determined. A difficulty arises when one spark is formed before the others. In this event, all the available charge may flow through this first spark, with the result that no sparks will be seen to correspond with the other particles. Also, in cases where one HV plate serves two gaps, often only one gap may fire due to this effect.

Ideally, all the sparks would develop at the same speed. But spark growth is a statistical phenomenon; it is mainly decided by the growth of the critical avalanche. In practice, the experimenter must aim to reduce the fluctuations—if possible, to the extent that they will be small in comparison with the discharge time constant. Crudely, and neglecting transit times, for a 1 cm gap, 1 m² chamber: $C = \sim 1000$ pF and $RC \sim 10^{-9}$ sec, if R is to be $\sim 1\ \Omega$.

The spark-formation times may be much reduced by using neon and by increasing the applied electric field (Fischer and Zorn, 1961) (see Fig. 3.2.1.1). These authors (1962) believe the spark-formation time can be reduced to about 2 nsec if the field strength be 25 kV/cm and if a trace of argon be added to the neon to allow the Penning effect to work.

Other factors influencing the multiple-track efficiency are the resistance of the plates, the rise-time of the pulse and the gap spacing. Figure 3.5.3.1 shows the two-spark efficiency curves of Michael and Schluter (1963) from which it is seen that large gaps and short rise-times give the best results. The Townsend coefficient varies fiercely with the field, and hence it is obvious that

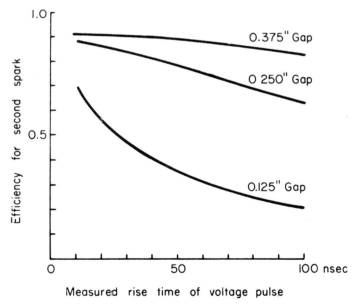

Fig. 3.5.3.1 Two-spark efficiency curves as a function of pulse rise-time. (Michael and Schluter, 1963)

uniform spark growths will be obtained only in very uniform gaps.

Faissner *et al.* (1963) studied the efficiencies for many tracks by observing electron showers in a system of 12 gaps of dimensions $100 \times 160 \times 1$ cm^3. Each HV plate was connected to a 12,000 pF capacitor charged to 12 kV. The cameras were set (f/8) to record the thin multiple tracks—i.e. single tracks were overexposed. The results (Fig. 3.5.3.2) show that the efficiency can remain high even where there are a large number of sparks in a gap. This result, however, probably depends upon the large separation of the sparks—even for ten sparks the average spacing is 10 cm.

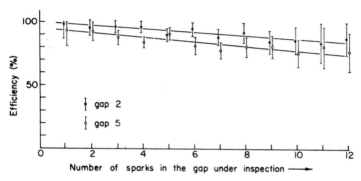

Fig. 3.5.3.2 Multitrack efficiencies in a large twelve-gap chamber. (Faissner *et al.*, 1963)

With the advent of wire chambers, higher efficiencies have resulted. The length and impedance of the wires implies that the sparks will be relatively isolated from each other, and hence robbing of energy is less likely to occur. This quality can be enhanced by connecting together the wires of a wire plane with a resistive strip (Galster *et al.*, 1966).

The ionizing power of the particle is of some relevance in a discussion of multitrack efficiency. The more primary electrons, the faster will be the creation of the initial avalanche. Thus, among several simultaneous particles, one might expect the most heavily ionizing to produce the earliest and hence brightest sparks. This is often observed, although the correlation is sufficiently unreliable to make it unattractive for use in measurements on ionization.

By considering the spark chamber as a transmission line, Allkofer *et al.* (1970) have obtained a theoretical expression for multitrack efficiencies. Their treatment included the manner in which sparks resulted in asymmetrical distortions of the applied field, and also the influence of reflexions from the boundaries of the chamber. The multitrack efficiency of a chamber could be improved by covering the electrodes with dielectrics—for this would effectively limit the spark current and hence the consequences of sparking.

3.5.4 Time resolution

Spark chambers often have to work in very intense beams—whose particle densities may reach 10^6 or more particles/second. From such a cataclysm, the pioneer may want to select just one particle with his counter/logic system. Clearly, it is necessary to preclude sparks due to the unwanted trajectories. This may be done by shortening the memory time of the chamber, e.g. in the above case, to about 1 μsec. If after this time, electrons from undesired events have disappeared, then a HV pulse—applied with a delay of, say, 200 nsec—has a good chance of initiating only the correct spark.

A good time resolution may be achieved by the application of a clearing field, and also by the addition of some suitable quenching gas. In fact, it is common to employ both techniques together.

The investigations of Culligan *et al.* (1961) showed that by increasing the clearing field from nothing to 80 V/cm across a 4 mm gap containing 99·99% pure argon, the memory time diminished from more than 30 μsec to about 400 nsec (see Fig. 3.5.4.1(*a*)). Nevertheless, it was noticed that a small efficiency of a few per cent remained for much longer times. It is thought that although the d.c. field will remove the free electrons, some electrons may be captured into loosely bound states, and it is these that are loosened when the HV field is applied.

The same studies showed that the addition of alcohol eliminates this nasty residual efficiency (Fig. 3.5.4.1(*b*)). Presumably, the alcohol molecules grasp the miscreant electrons. Fischer's

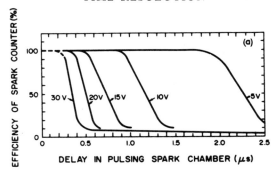

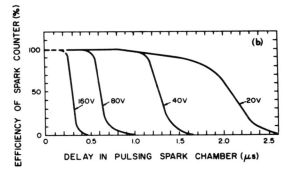

Fig. 3.5.4.1 Efficiency of a 4 mm gap as a function of delay and clearing field when filled with (a) argon; (b) argon saturated with alcohol. (Culligan et al., 1961)

(1966) studies on a conventional gas—neon + 10% helium + 1·5% ethyl alcohol—showed that resolving times down to 200 ns are obtainable. In fact, for most experiments this is quite good enough—for the delay in triggering is of this order!

Ideally, one would like a vertical cut-off after a prescribed time, in the efficiency curve. Fischer's curves showed how closely this ideal is approached. They also confirmed other findings (see section 3.5.1) that the resolving time does not decrease monotonically with the clearing field—the same curve applies to 200 and 450 V. Some uncertainty exists as to its cause (Schneider and Höhne, 1963).

A number of studies have been made on the effects of gas composition. Schneider and Höhne, for example, find that the addition of 1% N_2 to argon ensures that the efficiency does not linger, but drops rapidly to zero.

Murphy and O'Neill (1962) have discussed the desirable qualities for a chemical quenching agent; in the light of the facts that the electrons, produced by ionization with energies of several electron volts, are thermalized (0·025 eV) in about 500 nsec.

Ideally, the additive should have a low cross-section for electrons above thermal velocities—so that they will not be removed before the HV pulse is applied—but a high cross-section for thermalized electrons. Most gases have capture cross-sections that go to zero at low energies since their attachment is due to endothermic chemical reactions. However, there is a small class of gases that captures by exothermic processes, and these have high capture probabilities for thermal electrons. The class includes NO, HCl, Cl_2 and SO_2 (Bradbury and Tatel, 1934). Murphy and O'Neill added $\frac{1}{2}$% SO_2 to argon and found the efficiency would drop to 10% in 2 μsec (with no clearing field).

3.5.5 Recovery time

The recovery time is the time that must elapse after a spark discharge before the chamber can usefully be pulsed again. During this time all the electrons, ions and excited atoms resulting from the spark plasma must be eliminated, so that no trace of the old track remains.

The processes of recovery have been studied by Fukui and Miyamoto (1961) and Autonès et al. (1963). In general, one can say that the dead-time will depend on the energy of the discharge, and on the value of the clearing field. Typically, dead-times might lie in the range 5–20 msec, but with the addition of a little alcohol vapour 1 msec is attainable.

The plasma channel that remains immediately the spark has passed is essentially neutral. After some recombination and ambipolar diffusion, the resolving-time clearing field becomes effective and rapidly drifts out the electrons. Slow positive ions are left together with some metastable atoms. To speed up the removal of the ions, Fischer (1965) has used a pulsed field of about 2 kV/cm. It is found that alcohol is useful to help quench the gas—noble ions undergo charge transfer with the alcohol molecules, and this precludes secondary electron emission on impact with the electrode. Figure 3.5.5.1 shows that with care the dead-time can be reduced to about 300 μsec.

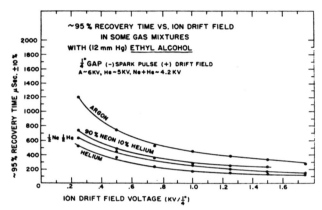

Fig. 3.5.5.1 Recovery times in a narrow-gap chamber. (Fischer, 1965)

3.5.6 Performance in magnetic fields

In some experiments it is desirable to insert sampling chambers in strong magnetic fields. This has been done successfully, but there are one or two effects that have to be taken into account.

The first is that the effective rate of electron diffusion across magnetic lines is considerably reduced. For example, Heyn (1961) has found that the diffusion distance in neon is reduced by a factor of two in a field of 15 kG. One significant result of this is to slow the action of the clearing field in emptying the gap of electrons.

A serious consequence of the combined influence of the magnetic field and the d.c. clearing field is that the electrons are caused to drift in the $E \times B$ direction. [The combination of the drifting and collision processes has been discussed by Townsend (1947).] In a many-gap array the clearing field normally alternates from one gap to the next, and so the drifting alternates too. As a result the sparks are staggered as is shown in Fig. 3.5.6.1.

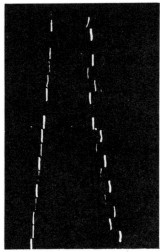

Fig. 3.5.6.1 Two orthogonal views of a track in a magnetic field. In the right-hand view, the magnetic field is along the line of sight. (Overseth, 1964)

The relative amounts of drifting as a function of clearing field, have been measured by Overseth (1964) in a magnetic field of 9 kG and with an HV pulse delay of 200 nsec.

In any experiment, if there be too much staggering, problems may arise in distinguishing tracks. Thus Astbury et al. (1967) have had to compromise, and reduce their clearing field from a preferred value of 200 V/cm down to 60 V/cm. Even so, the alternate sparks are displaced by 1·1 mm to either side of the true trajectory. Another attempt to solve the problem was indicated in section 3.5.4, i.e. to dispense with the d.c. field and clear chemically. Murphy and O'Neill (1962) tried SO_2.

TABLE 3.5.6 (Overseth, 1964)
Relative $E \times B$ drift for various gases

Gas	Clearing field V/cm			
	0	40	100	200
Helium	0·3	1·2	1·3	1·4
90% Neon + 10% Helium	0·4	2·2	5·0	—
Argon	0·6	2·6	7·0	—

3.5.7 Performance at high pressures

Operation at high pressures offers the prospect of a high spatial resolution. Narrower gaps would be possible and thus inaccuracies due to the spark not following the track inclination, and to spark instabilities, would be reduced. And because the diffusion is inversely proportional to the square root of the pressure, the width of the sparks would be improved too.

With a prototype optical chamber lying within a pressure vessel capable of withstanding 40 atm, Willis *et al.* (1971) obtained a resolution of 70 μm at 15 atm. The filling was 90% neon/10% helium and the aluminium electrode spacing 3 mm. The 22 kV HV pulse was derived from a spark-gap pulser with a 5000 pF storage capacitor. A good multitrack efficiency was obtained.

As a result of these findings Willis *et al.* (1971) proceeded to construct a wire-plane chamber capable of exploiting the

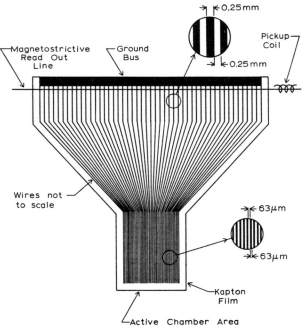

Fig. 3.5.7.1 Schematic view of a wire plane for use in a high-pressure chamber. In fact there are 300 wires. (Willis *et al.*, 1971)

advantages. Figure 3.5.7.1 shows a 4 × 4 cm plane with 8 wires/mm fabricated by photoetching 10 μm copper on a backing of 125 μm Kapton film. A gap size of 1·2 mm was created by a 150 μm beryllium second electrode. The windows of the chamber were stainless-steel burster discs calibrated to burst at 35 atm. Voltages of 7 kV were suitable for a neon/helium filling at 15 atm. Care had to be taken to ensure the rise-times were not above 20 ns. Single-track efficiencies of up to 99% were attainable if a little quenching gas was added. Figure 3.5.7.2 shows some efficiency curves and indicates the remarkably low clearing fields required under these conditions.

Magnetostriction was employed for digitizing the sparking location, and this was facilitated by the wire fan out shown in the diagram. The overall chamber resolution was estimated to be 75 μm—which is considerably better than the 0·2 mm "good value" of conventional spark chambers.

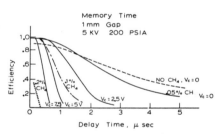

Fig. 3.5.7.2 Efficiency curves in a high-pressure, very-narrow-gap chamber. The CH_4 was accompanied by ten times as much argon in each case. (Willis *et al.*, 1971)

CHAPTER 4

Measurement in Sampling Spark Chambers

MANY methods have been used to record the information indicated by the sparks. The techniques described in the following sections are all practical, but it may be noted that most sophisticated projects have favoured immediate digitization with wire planes (see also section 3.3). Fischer (1966), for example, has listed forty high-energy experiments, half of which use core read-out and the other half magnetostriction.

The requirements of the experiment and the equipment available are the criteria for the choice of method, e.g. the necessary cycling rate; the required accuracy; whether many simultaneous tracks must be recorded; whether photography is possible; whether on-line computers are available; whether magnetic fields are present, etc., to say nothing of the cost.

4.1 RECORDING WITH PHOTOGRAPHIC FILM

Photographic film was used to record the sparks in almost all the early experiments with spark chambers. This method has several virtues: the photography can be relatively simple; the cameras can be operated in magnetic fields; the negative is a permanent record; and the information is in a form that can be studied by that marvellous decision-making system—the human eye.

On the other hand, the technique is very slow, requiring subsequent development of the film, etc., and it therefore cannot give immediate results to allow adjustment of the experiment. The rate of taking pictures is often limited by the mechanical operation of the shutter and film transporter. Experiments which involve large numbers of events need masses of film and require analysis with a system such as H.P.D., i.e. they incur many of the problems—and expense—of bubble-chamber analysis. The method also has the disadvantage that bright sparks are necessary, whereas it is advantageous at high cycling rates (10^4/sec) to use sparks with less energy.

4.1.1 Optics

The simplest arrangement is merely to place a standard 35 mm camera some distance away from the chamber. A common panchromatic film (e.g. Ilford HP4, 400 a.s.a.) will be quite sufficient to register sparks with a small camera aperture. The shutter will have to be opened prior to the triggering of the chamber. With an elementary chamber a three-dimensional record may be obtained with two cameras mounted at 90°. Alternatively, one can acquire two views in one frame by mounting a plane mirror to one side and photographing the sparks directly, and also their reflections.

For most experiments, however, the configuration of spark gaps requires a system of lenses, prisms or mirrors, to allow accurate measurements to be made. One technique is to use large spherical or cylindrical lenses to extract the light from the parallel gaps (Fig. 4.1.1.1(a)). For example, large spherical lenses have been made from perspex with a diameter of 4 ft and a focal length of 15 ft (Cronin, 1967). Care has to be taken to reduce spherical aberration, e.g. an f/4 aperture results in a 2% displacement of the focal point. Spherical lenses are preferable to cylindrical; the latter produce a distorted image that has to be corrected for mathematically.

An alternative is the use of long prisms whose refracting surfaces are independently cut to suitable angles (Fig. 4.1.1.1(b)). A good example of their use is the design by Astbury *et al.* (1967) of a 78-gap chamber inside a large volume magnetic field. In Fig. 4.1.1.2 it is seen that in the plane of the aluminium foil electrodes the stereo views (on 70 mm film) are taken by reflection in mirrors about 50 cm apart. In the plane perpendicular to the foils, the spark light from the narrow gaps is refracted to the camera plane by precisely machined and polished perspex prisms. After a path of 270 cm, the light from the two views passed through two objectives (10·8 cm focal length) to be recorded at a demagnification of 24·2. No problems concerning depth of field arose.

Mirrors (Fig. 4.1.1.1(c)) have been used in some investi-

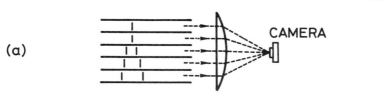

(a)

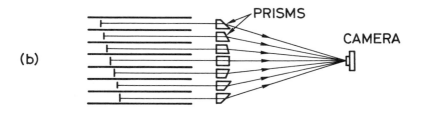

(b)

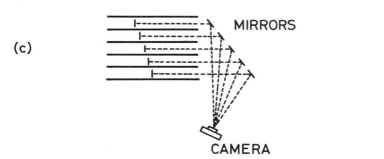

(c)

Fig. 4.1.1.1 Photographic recording of sparks: (*a*) with a large spherical or cylindrical lens; (*b*) with long independently cut prisms; (*c*) with mirrors.

gations: an example is the neutrino experiment of Faissner *et al.* (1963) in which 6 ft mirrors were used.

It is inevitable that in optical arrangements employing mirrors and prisms there will be astigmatism, and in large systems significant distortions due to imperfections are likely to arise. It is thus most important that the photographs should contain many fiducial marks, accurately to locate the images. Uniform fiducials can be obtained by sending light pulses from triggered xenon

132 MEASUREMENT IN SAMPLING SPARK CHAMBERS

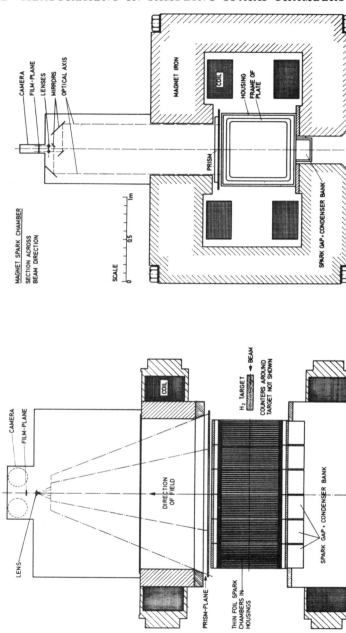

Fig. 4.1.1.2 Photographic recording, using prisms, for large spark-chamber system mounted within a magnetic field. Stereo views are obtained with the aid of mirrors. (Astbury et al., 1967)

flashtubes down light guides, momentarily to illuminate the crosses. The whole field may then be calibrated by photographing fine wires whose position is precisely known. It is customary to include a luminous datum box so that the event and the operating conditions can be identified in the film.

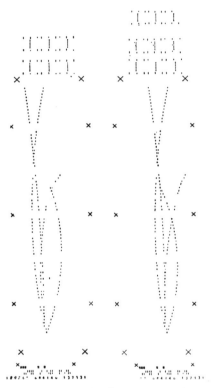

Fig. 4.1.1.3 Stereo photograph of an event taken with the CERN magnet spark chamber. (Fig. 4.1.1.2; Astbury *et al.*, 1964)

A number of unorthodox solutions to the optical problem have been reported. In a hodoscope arrangement of many gaps by Banaigs *et al.* (1964), diaphanous wire-mesh electrodes were employed, with the result that photographs could be taken parallel to the electric field. The particle tracks then appeared as rows of dots, with each dot being associated with a specific gap.

Another possibility is to use very shiny foils as electrodes, so that the light from sparks within the narrow gaps may undergo multiple reflection in the foils before emerging and impinging on the camera. The result is similar to that obtained with a cylindrical lens. In an adaptation of this method, Wenzel (1964), divided a cylindrical multigap chamber into two semi-cylinders. Using one-half as a chamber, the radial and azimuthal positions of the sparks could be measured by focusing directly through a spherical lens at one end. The longitudinal position could be detected with a camera mounted perpendicularly to the bisected edge, the image being formed by light that escaped after undergoing multiple reflection in the curved electrodes.

4.1.2 Photography

The photography of sparks has not been difficult. Good results may be achieved with a typical panchromatic film (say 400 a.s.a): 0·04 joule sparks will be nicely detected at f/11.

The variables in spark-chamber photography are the quality and speed of the film, the camera aperture, the spark energy and the demagnification. To allow a depth of focus large enough to cover the dimension of a typical large chamber ($\sim$ 1–2 metres), ones requires small apertures (e.g. f/20). To ensure thin sparks, small spark energies (hence intensities) must be used. These suggest fast film. The greater the demagnification, the better must be the emulsion. Cronin (1967) has said that for a precision of 0·01 in., at a demagnification of 100, a film accuracy of 2·5 μ is required. This is close to the limit.

It is beyond the function of this book to enter into details of cameras and the film transport mechanisms. Suffice it to say that Welford (1967) has written a useful article on this topic, and Cronin (1967) has mentioned that 35 mm cameras with cycling times of 50 msec, and even down to 10 msec, are commercially available. This latter time is, of course, of the same order as many spark-chamber recovery times.

4.1.3 Digitization and analysis

Where possible, the multitudinous film is nowadays processed, measured and analysed automatically. It helped enormously that bubble-chamber film systems had been developed earlier—

DIGITIZATION AND ANALYSIS

for it was easy to adapt the simpler format of spark photographs (geometrically symmetrical gaps, etc.). Even moderately complex pictures can now be handled at rates down to one per second—and the scanning girls are fast disappearing.

The analysis is accomplished in a number of steps. First, the spark coordinates (and sometimes, intensities, widths, etc.) may be found with a flying-spot digitizer [or H.P.D. after Hough and Powell (Hough, 1967)] (Fig. 4.1.3.1). A fine pencil beam of light

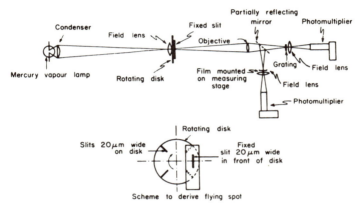

Fig. 4.1.3.1 The Hough-Powell Digitizer for extracting the positions of sparks recorded on film. A flying spot of light scans the film and a grating. One photomultiplier detects the sparks, while the output of the other (behind the grating) records the corresponding instantaneous positions of the spot. (Hough, 1967)

under the control of a computer scans the photograph in a raster fashion; and a photomultiplier emits pulses corresponding to changes in the intensity of transmitted light as the synchronized beam traverses a spark. Electronics send the spark addresses (in binary form) to the on-line computer. It is convenient to scan each gap, say, three times.

Prior to determining the spark coordinates on a particular frame, the computer first has to recognize the fiducial marks (big X). Many traversals of the 45° branches of the X's allow their exact centres to be computed. The identification of the sparks in real space may then proceed. With suitable programming, the computer can search for certain events (e.g. Fig.

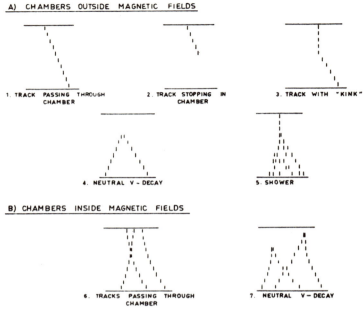

Fig. 4.1.3.2 Examples of track patterns that may be automatically recognized by a computer. (Zanella, 1968)

4.1.3.2), reconstruct them geometrically, perform a kinematical analysis, and provide a statistical evaluation of the experimental data (Zanella, 1968).

4.1.4 Precision

In a many-gap system it is possible to construct a good fit for a particle trajectory and from this calculate the individual spark deviations. Any estimate of the error in determination of individual spark coordinates from film will have contributions from spark formation, imperfect optics and the digitizing.

Astbury *et al.* (1967) considered that their spark fluctuations (rms) were ± 0.18 mm and their measurement ($+$ optic) error ± 0.09 mm. They pointed out that the 0.09 mm in space corresponded to 3.7 μm on the film—which is bigger than the size of the flying spot (1.6 μm) and much smaller than the width of the spark (80 μm). (N.B. each spark was scanned three times.)

In another assessment of precision, Christenson *et al.* (1964) obtained 0·21 mm for the spark deviation when the tracks were perpendicular to the plates, but this increased to 0·44 mm for angles of 23°.

4.2 SONIC DETECTION

4.2.1 Principles of measurement

One of the earliest filmless recording systems used acoustic detection (e.g. Bardon *et al.*, 1964). When a spark occurs between two parallel-plane electrodes, a shock-wave is created at the position of the spark. The circular wavefront travels outwards with the velocity of sound in the specific gas. The location of the spark may be determined by measuring the time intervals occurring between the formation of the spark and the arrival of the waves at a number of judiciously placed microphones.

The simplest case is just two microphones—say A and B in Fig. 4.2.1.1(*a*). A spark at S would be determined by the times for the sound-wave to reach the microphones at A and B; the time intervals being measured (i.e. digitized) in terms of the number of pulses from a high-frequency oscillator, say, 20 MHz recorded in a scaler.

The speed of sound in a specific gas at atmospheric pressure depends upon the gas and its temperature. At 0°C it is 435 m/sec in neon, 319 m/sec in argon, 965 m/sec in helium and 1284 m/sec

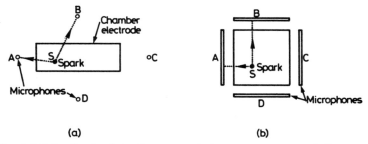

(a) (b)

Fig. 4.2.1.1 Sonic detection: travel times are measured for spark sound waves to arrive at the microphone : (*a*) with four-point microphones; (*b*) with four long microphones; these give Cartesian coordinates directly.

in hydrogen. In the spark chamber mixture of 90% neon and 10% helium it is about 500 m/sec. As the temperature may easily vary over the course of an experiment, the velocity must be continually monitored with the aid of a test spark.

In practice, the situation is a little complicated. In particular, the velocity of the pressure wave is not exactly constant: the velocity at a given instant is a function of the distance the wave has travelled. As the distance increases so the velocity approximates more closely to the velocity of sound. Lillethun et al. (1964) showed that the effect of this variation was equivalent to assigning to the spark a finite size ΔR (typically about 4 mm). The distance from the spark to the microphone is then given by $R = v_0 t + \Delta R$ where v_0 is the asymptotic velocity. ΔR is a function of the spark energy.

The two coordinates of one spark may be ascertained with just two microphones—by measuring the times and assuming a value for the sound velocity. Lillethun et al. employed four microphones (Fig. 4.2.1.1(a)) with the advantage that the velocity may be eliminated from the equations, and the coordinates then depend only on the measured times. With a computer, they were able to locate a spark with a precision of ± 0.2 mm in a chamber of 50×15 cm^2.

4.2.2 Examples of sonic chambers

The sonic chambers of Bardon et al. (1964) consisted of $30 \times 30 \times 0.0025$ cm^3 aluminium foil electrodes separated by a 6 mm gap. A piezo-electric probe was mounted at each corner. They were lead-zirconate cylindrical shells, 12 mm long and 3 mm outer diameter, whose inner and outer surfaces were silver-plated. Each probe was soldered to a coaxial cable and held in position by shock-absorbent rubber "O" rings. The transducers produced a pulse of several millivolts when a 0·04 joule spark occurred. The pulse had a rise-time of about 1 μsec and was followed by ringing for a few milliseconds. The leading edge was used in the time measurement, and because the acoustic wave amplitude diminishes as $R^{-3/2}$ in traversing a distance R, it was arranged that the electronic threshold for closing the scalers decreased with time to compensate.

One feature of the electronics was that the scalers were only switched on to count the oscillator pulses at the end of a precise interval (60 μsec) after the spark discharge. This allowed all electrical noise to die down before the counting began. Jones et al. (1964) have described the use of long microphones extending the whole length of a gap (Fig. 4.2.1.1(b)). The main advantage is that the x and y coordinates are obtained directly, thus reducing the need for computation. Two types of microphone were successfully employed (Fig. 4.2.2.1). Electrostatic air-gap condensers were constructed from 110 cm lengths of aluminium alloy bar, milled and fabricated with a thin aluminized melinex electrode. Long piezo-electric detectors were also made from 7·5 cm lengths of lead–zirconate–titanate crystal that had been silvered on opposite surfaces. When care had been taken in the alignment of the microphones an accuracy of 0·3 mm was obtained.

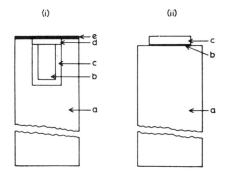

(i) Section through electrostatic microphone
(a) Aluminium alloy bar of 0.6 × 3 cm² cross section with a slot of 0.4 × 0.6 cm² cross section milled into it.
(b) Alloy strip 0.2 × 0.5 cm².
(c) Araldite (MY 753 + HY 951) to support and insulate (b).
(d) Air gap of thickness 0.01 cm.
(e) Aluminized mylar diaphragm. The mylar was 6 μm thick; the aluminized layer 50 μmm.

(ii) Section through piezoelectric microphone
(a) Brass bar 0.6 × 2.5 cm².
(b) Conducting araldite.
(c) Crystal of lead zirconate titanate 0.4 × 0.1 cm² silvered on upper and lower faces.

Fig. 4.2.2.1 Details of two types of long microphone. (Jones et al., 1964)

A variation on the sonic theme has been proposed in which the sound-wave travels down an electrode, rather than in the gas. Piezo-electric crystals, attached at the edges of the electrodes, detected the spark vibrations and a fast acquisition of data was claimed (Brinckmann, 1969).

4.2.3 Advantages and difficulties

In recent years sonic chambers have not been as fashionable as wire-plane chambers. However, they have the virtue that they can be operated in a magnetic field, and in general this is where they are to be found.

Their main disadvantage is the limited number of simultaneous sparks they can record. Even two sparks present considerable analytic difficulty, e.g. Lillethun and Zanella (1964) suggest that about eight microphones would be necessary to pinpoint separate sparks. Apart from the mathematics, there are the experimental limitations: the sound-waves reflected off the chamber walls; the long ringing times and the necessity to damp the microphones; and in the case of long transducers, the (dead) time for the sound-wave to traverse them.

Another disadvantage is the considerable variation in the velocity of sound with temperature. It is given by

$$\frac{v(T)}{v(273)} = \frac{T}{273}.$$

Thus a change of 1°K alters the velocity by 0·2%. This means that in a one-metre path length the measured distance would be out by 2 mm, if no corrections were made.

Nevertheless, Maglić (1966) has, with spirit and nice humour, pointed out that sonic chambers are simple to make; their recovery times can be reduced to less than 1 msec; the resolution of a two-metre chamber can be 0·5 mm; the read-out time is of the order of 10^{-5} sec; and that even three sparks in a gap could be handled.

4.3 TELEVISION VIDICON CAMERA RECORDING

The use of a television camera automatically to record sparks was first proposed by Gelertner in 1961. It is the only filmless method to use the optical emission of sparks, and it has been used successfully to digitize spark addresses in a number of experiments. It is quite possible that further development of the camera tubes will lead to their becoming very popular in the future.

4.3.1 Explanation of vidicon

The diagram (Fig. 4.3.1,1) illustrates the operation of the vidicon tube. The chamber scene is focused through the glass wall, through a transparent electrode and on to a photoconductive

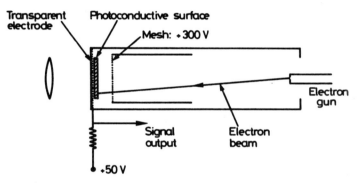

Fig. 4.3.1.1 Schematic view of a vidicon television camera.

surface deposited inside. Typically, this layer is antimony trisulphide. Within the tube a fine electron beam emanating from the gun (and driven by suitable circuits) scans the photoconductive surface in a regular raster fashion with the effect that a steady voltage is normally (in the dark) maintained across it. If the image of a spark falls on this surface, the local capacitance at this point discharges. When the spark has died away, the surface reverts to its insulating state, so that when the electron beam next scans the point, the voltage is restored in the local capacitance and a current pulse is observed in the resistor. An oustanding virtue of this common camera tube is the high density of information that can be stored on its photosensitive face.

4.3.2 Application to spark chambers

The manner of raster scanning is a matter of choice; broadcasting might use 625 lines, but the number of scans for spark-chamber viewing can be suited to the experiment. The simplest case is to arrange that one horizontal sweep of the electron beam be made to correspond with the image of each gap (e.g. Crawford et al., 1967).

The beam is synchronized with a 20 MHz oscillator. A number of scalers start to count the oscillations when the sweep starts to scan. The signal pulse caused by the first spark image to be encountered closes the first scaler; the stored number of oscillations is then a measure of the position of the first spark. The remaining scalers continue to count, only to close in turn, to yield the locations of successive sparks in the same sweep. The numbers are immediately transferred to a buffer store, and the system proceeds to the next gap.

The method often requires an optical system to bring the light to the vidicon face. A picture of a Chicago system is given in Fig. 4.3.2.1 (Hincks et al., 1966). Stereoscopic (90°) views are recorded in one camera with a pertinent arrangement of mirrors. Alternatively, mirrors may be dispensed with, and large systems of separated chambers observed directly with many cameras.

The height of a video pulse will vary with the luminosity of the spark—and this will depend on the number of other sparks in the chamber. To ensure reliable estimation of the spark centroid regardless of height, the zero-crossing technique (Fig. 4.5.4.1) is used.

The method of digitizing the spark position with oscillator pulses implies that each pulse is equivalent to a distance in the chamber. In the Chicago experiment, where each gap was scanned five times, one 10 MHz pulse corresponded to 0·5 mm in real space.

After recording, the beam had to scan a couple of times to erase the previous event. The time to scan the 525-line vidicons viewing sixteen chambers was 33 msec, and therefore the total time to read one event was about 99 msec. Krienen (1969) suggests that a rate of thirty per second would be feasible for 735 lines with a clock frequency of 40 MHz. Faster rates are obtainable by reducing the lines scanned by each vidicon and relying on only one erasing scan. Anderson (1964) suggests that by sweeping only ten lines, the total event recording time can be reduced to 1·5 msec—which is of the order of spark-chamber recovery times. Of course to operate with one erase implies that the discriminator must not only cut off the noise but the old tracks too, and this may lead to losing faint sparks.

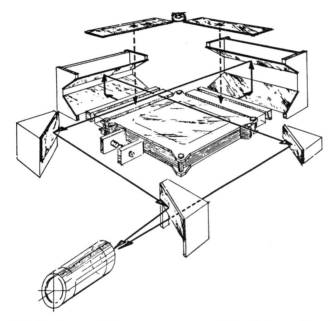

Fig. 4.3.2.1 An optical arrangement used in Chicago for vidicon detection. (Hincks *et al.*, 1966)

4.3.3 Accuracy

The precision with which a spark location may be converted into a binary number is typically quoted as one part in a thousand, for one line scan. Hincks *et al.* (1966) estimated their precision to be 0·3 mm in 25 cm chambers. A better figure of 1 in 4000 has been claimed by Lipman *et al.* (1969). If each gap is scanned many times, or if many gaps be used to define a trajectory, the precision can be improved. The minimum distance between two resolved sparks (two-spark resolution) is normally about five times the values for the precision.

The fact that the precision of a vidicon is quoted as a ratio of the full sweep, imposes a limit on the size of the chamber that may be viewed. If an optical chamber has an inherent setting error of 0·3 mm, the area viewed by a vidicon of 1/1000 precision can be up to 30 cm without loss of accuracy. A precision of 1/1000, as Terwilliger (1966) has pointed out, means that the accuracy for a frame of a vidicon is down by a factor of about eight on a 35 mm film system. A number of vidicons may be used to extend the measurable scene.

The method suffers from some disadvantages. The standard tubes may not be as sensitive as film to faint sparks. In comparison with wire chambers, there is the necessity to have an optical path. Although the spark chambers may exist inside strong magnetic fields, the vidicons themselves must be placed outside. A rather serious feature is that vidicons tend to be very sensitive to temperature; for example, Crawford et al. (1967) quote a drift of $0·7\%/°C$. The use of fiducials, however, may reduce the effects of this sensitivity. And, lastly, they are not as fast as experimenters would like, partly because of the need for erasing scans.

4.3.4 Rosy prospects for vidicons

A number of optimistic assessments have been made of the prospects for vidicons. They depend on the rapid technical developments that gain impetus from the television industry. As the tubes become faster and more sensitive, so their usefulness will become more obvious.

Krienen (1969) has referred to the Mullard "plumbicon". This device has a lead monoxide photoconductive target, rather than the antimony trisulphide of the vidicon. It has a much lower noise level, and a smaller time-lag which implies that one scan will be sufficient to erase the spark-chamber image to such a low level ($<15\%$) that a new event could then be admitted.

In another report, Roberts (1969) has pointed out that RCA tubes with 4000- and 5000-line resolutions are now available. (N.B. a 5000-line tube will give a 0·3 mm resolution over a 1·5 metre square field.) New high-sensitivity silicon diode vidicons are being produced which should prove adequate for recording streamer chamber tracks. And in some tubes the erasing scan has been eliminated. Roberts argues that in future very large systems will be more economically digitized with vidicons than wire-arrays. He mentions a SLAC proposal for a forty-eight wire-plane chamber system that together with the interface electronics, will cost $850,000. He suggests that wire-array costs rise more rapidly with size of chamber and complexity of event, than do vidicon costs.

An account of a plumbicon television camera system working successfully on the OMEGA spectrometer at CERN was given at Frascati, 1973 (Garvey et al.). With an apparatus of 100 thin foil spark chamber gaps plus six cameras, the following parameters were quoted: accuracy, 0·5 mm; two-spark resolution, 9 mm; dead time, 20 ms; sensitive time, $1\mu s$; six-track capability, 70%; cost, 500,000 Swiss Francs.

4.4 MAGNETIC CORE READ-OUT

This method rests on the properties and consequences of wire planes; and these were described in section 3.3.

4.4.1 Ferrite cores

The tiny ferrite cores used in computer memories have square hysteresis curves. If a core be threaded by a wire through which a current pulse passes, the magnetization may be driven from one stable state to the other. This is a basis for spark coordinate determination using wire planes with each wire in the plane threading its own core. It is an advantage that the state of the cores may be read with conventional computer electronics.

Figure 4.4.1.1 demonstrates the principles of the method. Initially, prior to the triggering of the chamber, all the cores are magnetized in such a direction that, when a spark does occur, the current in the participating wire reverses the magnetization of the appropriate core. Read and sense wires also thread the cores so that after the event their states may be rapidly determined in the same fashion as in digital computers. It is clear that this method presents no obvious limit to the number of simultaneous sparks which may be recorded.

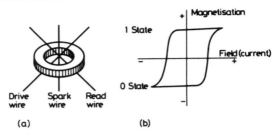
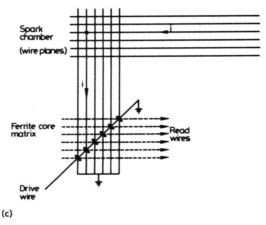

Fig. 4.4.1.1 Magnetic cores with wire planes: (*a*) one ferrite core; (*b*) the "square" hysteresis curve for ferrite; (*c*) crossed wire planes; only the core through which the spark current passes will change its magnetization.

4.4.2 Examples of chambers

It was Krienen (1963) who initiated the technique. In his prototype chamber the gap was bounded by 1000 cm² electrodes: one a high-voltage plate, the other an earthed wire plane of 256 wires (0·15 mm diameter, 1·27 mm spacing). In the intervening years, the core technique has been so developed that large chambers composing thousands of wires can be interrogated in a few hundred microseconds.

An example is that of Fischer (1966). His planes consisted of grids of fine (0·015 × 0·5 mm) copper lines printed on 0·07 mm fibre-glass sheets, each line being matched with a threaded core. A total of 1600 cores had to be questioned.

4.4.3 Read-out

For a large number of cores the method of read-out is a matter of compromise. Simple electronics could read each core in turn, i.e. serially—but the whole process might take a time of the order of a millisecond, which would be too long. On the other hand to read 1000 cores simultaneously, i.e. in parallel—would be most complicated.

For Fischer's chambers, Higinbotham *et al.* (1965) arranged the cores in 50 groups of 32 (Fig. 4.4.3.1). Each group was questioned in turn—the state of each core in the group being indicated by its own sense line organized in parallel read-out. The data were transferred to a buffer with a capacity of 4096 words (each containing 48 bits, i.e. three spark addresses). For events that contained up to ten sparks, the total read-out time was about 220 μsec, which was less than the chamber recovery time. All the data taken in the many spark-chamber firings

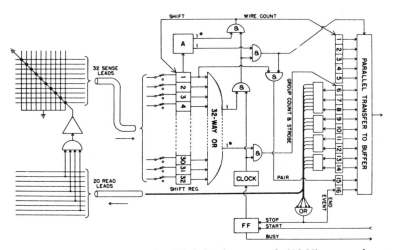

Fig. 4.4.3.1 The scheme by Higinbotham *et al.* (1965) to read out the information stored in a large number of ferrite cores. The cores are connected in groups of 32, and each group is read in succession.

during an accelerator pulse were stored in the buffer, and the contents were then read-out on to magnetic tape, or straight to the computer, between pulses.

4.4.4 Precision

For magnetic core chambers in general, the possible precision of measurement is governed mainly by the wire separation and the correlation of the particle track with the spark discharge. Typical accuracies are in the region 0·3 to 0·5 mm with wire spacings of the order of 0·6 to 1 mm. Galster *et al.* (1969), on passing 2·5 GeV electrons through four successive chambers, each constituted by two crossed wire planes, observed that the distributions of the spark positions about the mean values, showed a Gaussian shape. The FWHM was $\pm 0\cdot 2$ mm for the earthed electrode and $\pm 0\cdot 28$ mm for the high-voltage electrode. Two-spark resolution has been reported down to 2 mm (Fischer, 1967).

4.4.5 Difficulties

A number of difficulties may arise with ferrite core recording. One, that occurs especially when large capacity planes are used, is the problem of spark current overshoot which may cause a core to reset to its original state (Galster *et al.*, 1967). This may be overcome by limiting the spark current by means of resistive decoupling as described in the previous chapter. For this mode of detection, of course, the sparks need not be visible; they need only to set the cores reliably. Less than 1A may be satisfactory.

Another difficulty arises when the planes are being used in strong magnetic fields, because in fields of more than 50 to 100 gauss cores fail to operate. If the cores are to lie outside the field, long leads with a significant inductance would be required. If they are to exist in a matrix closely alongside the wire plane, then a large and an awkward amount of magnetic shielding would be necessary. Van Rossum (1968), at CERN, has instead collected all the cores together in a small box, which may be adequately shielded by a small amount of iron.

Fryberger *et al.* (1968) have commented that memory cores are more sensitive to magnetic fields that are perpendicular to their axis, than to fields along their axis—by a factor of 4 or 5. Advantage may therefore be gained by choosing an optimum orientation for the cores with respect to the magnetic field. In their application, at the best orientation, they surrounded their core boards with shield plates and found on doing so that the maximum ambient field strengths for satisfactory operation rose from 100; 77; 80 to 380; 460; 170 gauss along three mutually perpendicular axes.

4.5 MAGNETOSTRICTIVE READ-OUT

The magnetostrictive delay-line method has been used with great success in many high-energy experiments—for example, at CERN in the splitting of the A_2 meson (Benz et al., 1968). As in the case of ferrite cores, wire planes are used and the wires passing the spark current are determined.

4.5.1 Magnetostrictive delay-line

Magnetostriction is the phenomenon in which a change in the magnetization of a ferromagnetic material produces a change in the physical dimensions. It is a property of the anisotropy of the crystal structure, and different effects are seen in iron, nickel, cobalt and their various alloys. For example, iron crystals expand in the direction of magnetization and contract perpendicularly, while the opposite is true for cobalt. In addition, the effect in some cases changes sign: as the field is increased in iron so the expansion changes to contraction. The inverse, or Villari, effect also occurs: a deformation of the substance alters its magnetization.

A magnetostrictive delay-line is shown in Fig. 4.5.1.1(*a*). At each end, the magnetostrictive wire is encircled by a tiny pick-up coil. A current pulse in coil 1 will cause a local deformation in the wire and a longitudinal mechanical pulse will travel in both directions with a velocity given by $\sqrt{(E/\rho)}$, where E is the elastic modulus and ρ the density. This velocity in Fe–Co alloy (50%–50%) is 5290 m/sec at 20°C. On arrival at coil 2 the inverse effect alters the magnetization and a voltage is induced in the coil. The time interval between pulses in the two coils will

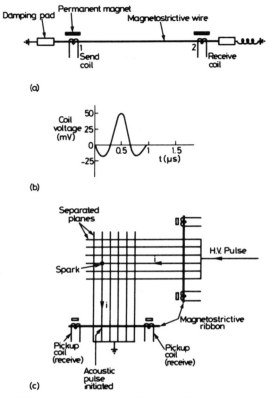

Fig. 4.5.1.1 Magnetostrictive read-out with wire planes: (*a*) An arrangement for testing the propagation of magnetostrictive pulses. (*b*) The shape of a typical receive coil pulse. (*c*) Magnetostrictive ribbons applied to two crossed wire planes. The spark current in the two wires initiates acoustic pulses in the ribbons, and their times of travel to the tiny pick-up coils are measured.

depend only on the distance between them. The wire must be damped at both ends to prevent reflections. The maximum effect is obtained when the magnetization inside the coils is adjusted to an optimum level by positioning small permanent magnets alongside.

Figure 4.5.1.1(*b*) shows a typical pulse detected in the pick-up coil. There is a clear peak that can operate the electronics.

4.5.2 Application to spark chambers

It was Gianelli (1964) who first applied magnetostriction to spark detection. In his experiment a magnetostrictive wire comprised the earthed electrode of a spark chamber. The spark to the wire, through its associated magnetic field, produced a local deformation whose longitudinal component travelled up the wire with very little attenuation. The delay between the spark and the pulse in the pick-up coil indicated the position along the wire where the spark occurred.

After a suggestion by Hine (1964), Perez-Mendez and Pfab (1965) initiated the technique that has since been adopted in many experiments. The method is illustrated in Fig. 4.5.1.1(c): the chamber electrodes are wire planes, i.e. they consist of hundreds of thin parallel wires that lie in a plane and are soldered to a common copper strap. Near each strap, a magnetostrictive ribbon lies at right angles over the wires but separated from them by very thin melinex tape. A spark current in any wire, flowing to the strap, passes so close to the ribbon that its field ($2i/4\pi\varepsilon_0 r$) creates a magnetostrictive pulse which travels in both directions along the ribbon, eventually to be detected by the tiny pick-up coils mounted at each end. Either of the two time-intervals indicates the position of the sparking wire. A useful check is also obtained since their sum should be constant—the time taken for a pulse to traverse the whole wire. The timing measurements are normally accomplished by counting 20 MHz oscillator pulses in suitably fast scalers—these being started at the moment of sparking and stopped by the pulse from a pick-up coil.

Any wire may initiate the magnetostrictive pulse and so the ribbon has to be magnetized along its whole length. Perez-Mendez found he could do this by passing a 25 A current pulse through the ribbon every time the chamber sparked. An alternative method is, once and for all, to stroke the ribbon with a magnet in such a direction that the field from the spark pulses reinforces the magnetization.

Although nickel ribbon* was used in the early work, an alloy (50% Fe, 50% Co) has come into general use. This expands in

---

* It appears there is less attenuation in ribbon than in circular section wire (Klanner, 1969).

the direction of an applied field. In a successful design due to Chikovani, Laverrière and Schübelin (1967) the ribbon was kept under tension along a rigid two-metre aluminium frame which was separate and demountable from the wire-plane electrodes. A mechanism was provided to adjust the tension and the ribbon passed through a 10 cm polythene tube containing silastic which damped the reflected mechanical pulse by a factor of 100. Tiny pick-up coils consisting of twenty turns of 0·06 mm copper wire were wound directly on the ribbon. The negligible mass of these coils meant that the magnetostrictive waves were not deformed. In operation the frames were fixed to the electrode supports with perspex clamps and the ribbon was insulated from the wires by 0·2 mm melinex sheet.

An alternative, static procedure has been described by Longo *et al.* (1971). Any sparking wire produces a localized patch of permanent magnetization in the magnetostrictive ribbon. The location of this patch may subsequently be determined by passing a current pulse down the ribbon, and observing the magnetostrictive signals in the usual way.

4.5.3 Digitization

Chikovani *et al.* were the first to digitize a wide gap (5 cm) chamber. In their application both electrodes were wire planes, and they were mounted at right angles to each other. In this way both x and y coordinates were obtained, and with several con-

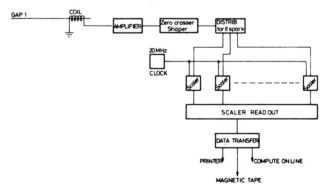

Fig. 4.5.3.1 A read-out system for the detection of up to eight sparks by magnetostrictive pick-up. (Chikovani, Laverriere and Schübelin, 1967)

tiguous chambers a complete identification of a trajectory was possible. A special difficulty arose in their case over the insulation of the read-out system from the 100 kV pulse which had to be supplied to the high-voltage side of the chamber. This was overcome by transmitting the output pulses through tiny ferrite isolating transformers.

The read-out system of Chikovani *et al.* is shown in Fig. 4.5.3.1. Up to eight simultaneous sparks could be accepted: the detector pulses being switched in turn to the scalers. The resolution of separate sparks is limited by the width of the acoustic pulse: Fig. 4.5.3.2 shows that sparks spaced by 4 mm are just resolved.

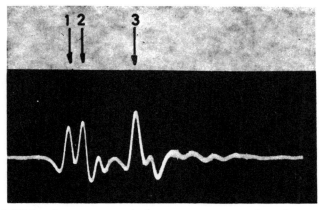

Fig. 4.5.3.2 A magnetostrictive output trace for three sparks. Sparks (1) and (2) have a separation of 4 mm and are just resolved. (Chikovani *et al.*, 1967)

4.5.4 Precision

The amplitude of the magnetostrictive output varies according to the number of sparks in the chamber. Therefore to avoid the errors associated with detecting the moment of arrival by a threshold voltage, the origin of the pulse is best indicated by the peak itself. A tunnel diode zero-crossing discriminator (Kirsten *et al.*, 1966) may be used. The pulse shown in Fig. 4.5.1.1(*b*) is first clipped—to remove the negative parts and produce a clean peak. It is then differentiated, and the moment the output voltage passes through zero corresponds to the arrival of the peak—regardless of amplitude (see Fig. 4.5.4.1). The zero-crossing technique is also helpful in giving the track position when many neighbouring wires participate in a single spark. The magnetostrictive signals from, say, three adjacent wires, are not resolved; rather, the magnetostrictive pulse is broadened.

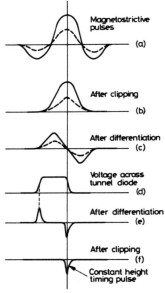

Fig. 4.5.4.1 The zero-crossing technique applied to magnetostrictive pulses. The output pulse corresponds in time with the peak of the input, and is independent of its amplitude.

Nevertheless, the zero-crossing discriminator will indicate the maximum, which is likely to be the track position.

The precision for determining a trajectory is limited not only by the nature of the acoustic pulse, but also by the geometry of the wire plane. The pulsed electric field will be distorted on approach to the planes; the field will concentrate on the wires and the spark will bend to follow it. In the everyday operation of 1 mm wire-spaced wide-gap chambers in the CERN boson spectrometer, an accuracy of 0·7 mm (FWHM) is obtained for nearly perpendicular trajectories, and about 1·5 mm for track inclinations in the region of 20° (Klanner, 1969). On the other hand, Grove et al. (1968) have reported an accuracy of 0·33 mm.

4.5.5 Difficulties due to multiple-sparking and to magnetic fields

One difficulty that has been encountered in the use of wire planes has appeared when many simultaneous sparks have had to be recorded. It is often found that the spark current in each wire suffers from "robbing", i.e. many sparks individually contain less current than one spark—with the result that, for example, whereas a pronounced magnetostrictive pulse is obtained for one spark, for say, 8 or 10, the pulses are barely above the noise level (Evans, 1969).

In addition, especially for crossed-wire planes, the spark efficiencies and currents over the surface vary because the inductance of the wires produces a non-uniform electric field in the gap when a high-voltage pulse is applied through the busbars of each plane. Grove et al. (1969) have alleviated these problems by incorporating the crossed-wire planes in a low-impedance transmission line (see section 3.4.3). By this means, the chamber field is made uniform and the source impedance is made small compared with the spark impedance. The voltage change on sparking is thus minimized, and the spark currents become less disparate.

Marshak and Pruss (1968) have indicated the difficulties in using magnetostrictive read-out in the presence of magnetic fields. It appears that the shape of the receive-coil output pulse is sensitive to the longitudinal field along the ribbon, the path travelled, the tension in the ribbon (Fig. 4.5.5.1), but not the transverse field. For example, a 200 gauss longitudinal field can reduce the pulse amplitude by about 50%, which renders it almost impossible for time measurements.

4.5.6 Use of the torsional mode of magnetostriction

It is known from acoustic theory, and also commonly observed, that longitudinal pulses suffer dispersive effects in their travel down a ribbon. A figure for the attenuation quoted at CERN is a factor of 2 in two metres. A practical limit is thus imposed on the feasible dimensions of chambers employing conventional magnetostrictive read-out.

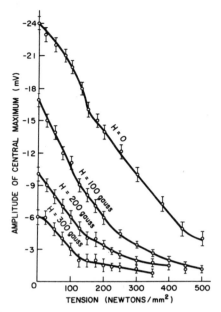

Fig. 4.5.5.1 A variation of the magnetostrictive pulse with tension and longitudinal magnetic field, in unannealed "Remendur". (Marshak and Pruss, 1968)

Table 4.5.6 indicates some properties of the longitudinal and torsional modes, and it may be seen that the dispersive effects may be minimized by choosing a sense wire diameter that is small compared to the mean effective wavelength of the pulse. Even with this condition satisfied (Fe–Co wire, 0·15 mm diameter) Grove et al. (1969) observed a 70% attenuation of the longitudinal pulse in 4 metres. They therefore suggested that large chambers should be read with torsional pulses since the lowest modes have no dispersion.

Torsional pulses may be generated by a combination of a steady-state azimuthal magnetic field through the wire and a local pulsed longitudinal field. The sum is a local pulsed helical field that generates a torsional pulse provided that $H_\theta > H_L$ (Fig. 4.5.6.1(a)). If the component H_θ is generated by a uniform current in the wire, it is a maximum at the surface and decreases to zero at the centre. In the region of the wire where $H_\theta < H_L$

USE OF TORSIONAL MODE OF MAGNETOSTRICTION

TABLE 4.5.6

Properties of longitudinal and torsional magnetostrictive pulses in wires
(Grove et al., 1969)

	Longitudinal	Torsional
Pulse generation	Joule effect	Wiedemann effect
Pulse detection	Inverse Joule (Viliari) effect	Inverse Weidemann effect
Longitudinal phase velocity in infinite medium	$C_L = (E_L/\rho)^{\frac{1}{2}} = \{(\lambda + 2\mu)/\rho\}^{\frac{1}{2}}$*	$C_T = (\mu/\rho)^{\frac{1}{2}}$ = velocity of transverse waves
Phase velocity V in wire of radius r, wavelength Λ	$V_L \approx C_L\{1 - (\pi^2\sigma^2 r^2/\Lambda^2)\}$†	$V_T = C_T$ (for lowest mode)
Wire radius r for useful pulse propagation	$r < 0\cdot 1$	$r < 5\cdot 14 V_T/(2\pi f_c)$ (for lowest mode)

* λ,μ = Lamé constants, E_L = Young's modulus of elasticity, ρ = density of material.
† $\sigma = \frac{1}{2}\lambda/(\lambda + \mu)$ = lat. contr./long. ext. = Poisson's ratio.

longitudinal pulses will be created. This simultaneous generation of longitudinal and torsional pulses poses no difficulty, for the detector may be constructed to detect one or the other.

Grove, Perez-Mendez and van Tuyl (1969) have detected the torsional pulses by converting them first to longitudinal pulses with the device shown in Fig. 4.5.6.1(b) and then observing the latter in the usual fashion. The device is made of a piece of nickel ribbon soldered to the sense wire. It is also possible to detect the torsional pulses in the orthodox arrangement (Fig. 4.5.1.1(a)) if no longitudinal field is present within the pick-up coils, i.e. if the biasing magnets are omitted. In this event, a voltage signal is induced in the coil by the inverse Wiedemann effect, i.e. by the time-varying helical magnetic field.

In practice various torsional modes may be excited, but the lowest can be selected by choosing the wire diameter so that it

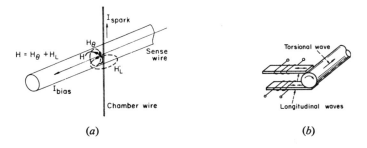

Fig. 4.5.6.1 (a) Generation, and (b) detection, of torsional magnetostrictive pulses. (Grove et al., 1969)

is the only one that suffers slight attenuation. The condition is that

$$r < 5.14 \frac{V_T}{2\pi f_c}$$

where $V_T = (\mu/\rho)^{\frac{1}{2}}$, and f_c is the limiting upper frequency for the lowest torsional mode. When this condition is satisfied for a given frequency f_c, the higher modes will be strongly attenuated by a factor

$$\alpha = \left(\frac{R_n}{r}\right)\left[1 - \left(\frac{f}{f_c}\right)^2\right]^{\frac{1}{2}}$$

where α is in neper/cm, and R_n are the roots of a characteristic equation and have values

$$R_1 = 5.14; \quad R_2 = 8.42; \quad R_3 = 11.62; \text{ etc.}$$

Grove et al. chose 0·02 in. wire of a hard drawn alloy of iron, nickel, chromium and titanium (Elvinar extra). The attenuation was found to be 40% in four metres. Although the pulses were slightly wider than longitudinal pulses, the two-spark resolution was unimpaired because the velocity is approximately two-thirds the velocity of the longitudinal pulses. They concluded that torsional pulse read-out would permit high accuracy of spark location along the entire length of a large chamber.

4.6 SPARKOSTRICTION

To overcome the deleterious effect of magnetic fields on magnetostrictive read-out, the Perez-Mendez team (Grove et al., 1968) has invented a method called "sparkostriction". The technique still employs the time of travel of an acoustic pulse in a suitable delay-line. However, reliance is placed not on magnetostriction, but on the generation of an acoustic pulse by a spark directly to the delay-line, and then on the subsequent detection of that pulse by a tiny piezo-electric crystal attached to the end of the delay-line.

The arrangement of electrical connections is such that a spark between the crossed-wire planes at Z (see Fig. 4.6.0.1(a)) is followed immediately by sparks at A and B, i.e. from the participating wires to the delay-lines. The delays before arrival of

SPARKOSTRICTION

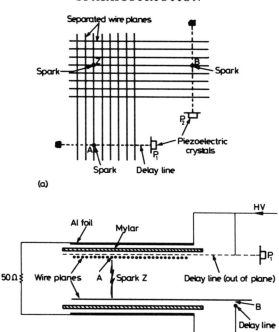

Fig. 4.6.0.1 Sparkostrictive read-out: (*a*) Spark at Z is immediately followed by sparks at A and B to the delay-lines. Acoustic pulses from A and B are timed on arrival at crystal detectors. (*b*) The electrical connection. (Grove *et al.*, 1968)

the mechanical pulses at the detectors P_1 and P_2 are measured with conventional 20 MHz pulses.

Figure 4.6.0.1(*b*) shows the arrangement by Grove *et al.* (1968). Their chamber was built to have transmission line characteristics: when the HV pulse is applied to the upper aluminium foil, a spark occurs (at the track location) between the wire planes, due to capacitative coupling. The fact that the delay-lines are connected to the HV and earth points, results in sparks occurring at A and B. The acoustic pulses were propagated down 0·0007 in. diameter BeCu alloy delay-line which was chosen for its conductivity, elasticity and good transmission properties. The pulses were detected with a piezo-electric transducer whose mount is shown in Fig. 4.6.0.2. It is seen that the crystal is electrically isolated from the sensor wire by the mylar film. Sparks, with a peak current of 10 A and total charge 3 μc, resulted in acoustic pulses that created, at the transducer output, a voltage amplitude of 10 mV and width 0·5 μsec.

It is clear that neither the acoustic propagation nor the piezoelectric detection will be affected by magnetic fields. Perez-Mendez, using a small (6 in. × 12 in.) chamber found that a spatial accuracy of ±0·4 mm was attainable, and so one might conclude that the method should considerably extend the application of wire planes in the presence of magnetic fields.

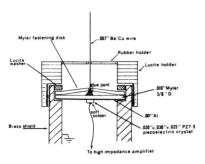

Fig. 4.6.0.2 Sparkostrictive transducer assembly. (Grove *et al.*, 1968)

4.7 CURRENT DISTRIBUTION METHOD

Charpak *et al.* (1963, 1966) have described a method for determining the position of a spark along a wire that is based on the relative magnitudes of the two currents that diverge from the spark. The charge (i.e. current) that is available for the spark divides into two according to the impedances to earth. If it be ensured that the effective impedances are solely those of the uniform wire, i.e. other connections negligible, then the difference in the currents will give the spark coordinate x. In Fig. 4.7.0.1

$$\text{Voltage at spark position} = xrI_1 = (l - x)rI_2$$

$$\text{and hence } I_1 - I_2 = (1 - 2x/l)I$$

CURRENT DISTRIBUTION METHOD

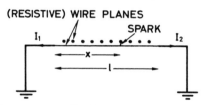

Fig. 4.7.0.1 The current distribution method. The spark current divides into I_1 and I_2 according to the lengths x and $l - x$

where r is the resistance per unit length of wire. The two currents are passed through separate core transformers, whose secondaries are connected in opposition: the position of the spark is then given by the height of the output pulse (e.g. 10 volts) which may be digitized in a conventional digital converter (Fig. 4.7.0.2).

In experiments, crossed-wire planes (e.g. 50 μm stainless steel,

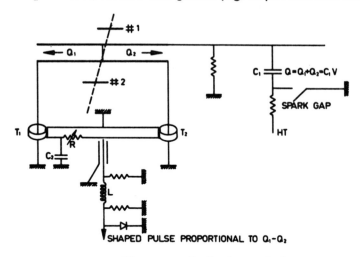

The current distribution method.
$Q = CV$ total charge delivered to the chamber.
Q_1 and Q_2 partial charges collected through the two ground leads
T_1 and T_2 transformer cores with winding in opposition.
R zero point adjusting resistance.

Fig. 4.7.0.2 The current distribution method. $I_1 - I_2$ determined from pulses out of transformers T_1 and T_2 with windings in opposition. (Charpak et al., 1965)

0·5 mm spacing) have been used so that both the x and y coordinates could be obtained—by observing currents in the HV side as well. Care has to be taken to ensure that all the wires (at both ends) connect properly with the copper connections.

From their and other work, Charpak *et al.* (1966) have concluded that the current distribution method is capable of giving a spark precision of ± 1 mm, although at the edges of crossed planes this figure increases to several mm.

Their main advantage is that the read-out is swift—within 200 ns after the event, which is much shorter than the memory time of most chamber gases. This is the reason that their usefulness appears largely to be as decision-making chambers, i.e. as elements of a complex system so that their quick recognition of a particle in a desired trajectory can be used to trigger conventional chambers.

The possible triggering rate is in the range 100 to 1000 per second; but a drawback is that only one particle can be assessed at a time.

The current-distribution method of read-out has been applied by Dahlgren *et al.* (1970) in a pair of coaxial cylindrical wire-spark chambers (Fig. 4.7.0.3). Their aim was to detect recoiling α particles after collisions between ^{4}He nuclei in the chamber and axial high-energy pions. To measure two coordinates—the

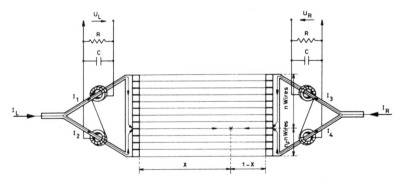

Fig. 4.7.0.3 Current distribution read-out from a wire in a cylindrical chamber. The inductance of the interconnections between the wires prevents the simple differences $I_R - I_L$ giving the longitudinal position (x) of the spark; it is necessary to introduce the branching shown, with two transformers at each end. (Dahlgren *et al.*, 1970)

longitudinal and azimuthal positions of the spark—each chamber consisted of cylindrical surfaces created by wires lying parallel to the axis, and having a 2 mm spacing. The HV surfaces were comprised of stainless-steel wires, of resistance 1 Ω/cm, which were joined at the ends by low-impedance interconnections. The HV pulses were fed into the wires from both ends, and in principle, the difference of the currents would give the longitudinal position of the spark. In practice, the effect of the (inductive) impedance of the circular end-connectors had to be reduced by feeding in the current at two points in each connector (Fig. 4.7.0.3). Thus four transformers were necessary, and a rather complicated expression gave the longitudinal position.

The earthed-wire surfaces were comprised of highly conductive, longitudinal, gold-covered molybdenum wires whose ends were interconnected with a circular constantan wire (1 Ω/cm). The azimuthal positions of a spark, i.e. the wire that sparked, was again determined by using the output pulses from transformers that were threaded by the current leads to the constantan connectors.

The two chambers had radii of 17 and 38 cm respectively. The angle of a recoil α particle track to the axis is given by the longitudinal coordinates of two sparks. A calibration with 5·48 MeV α indicated a measurement precision of 1·2° FWHM.

4.8 CAPACITATIVE AND OTHER MEMORIES

A new class of read-out techniques that is independent of magnetic fields, promises to be important in the future. The basic idea originated with Quercigh (1966), who suggested that the spark charge might be collected on a capacitor, which would then act as a memory. In the original work, each wire was connected to earth via its own 50,000 pF capacitor (Fig. 4.8.0.1(*a*)). After an event, the state of each capacitor was questioned in turn by a commutator that connected in an oscilloscope.

A more practical means for reading out the information might be to assess the capacitors sequentially through complementary sensing transistors (Schüller, 1969). In Fig. 4.8.0.1(*b*), diode D_1

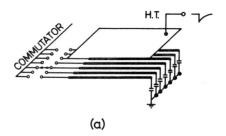

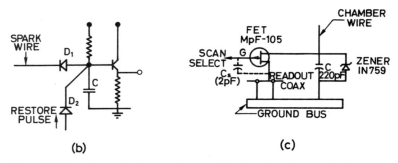

Fig. 4.8.0.1 Capacitative read-out: (a) Sketch of a spark chamber with a memory capacitor connected to each wire (Quercigh, 1966). (b) Reading the capacitor charge with a transistor (Schüller, 1969). (c) Reading the capacitor with an F.E.T. (Neumann and Nunamaker, 1968)

is to prevent the charge escaping back through the spark plasma, and D_2 is to enable a pulse, after assessment, to restore the original state of the memory capacitor.

In a sensitive extension, Neumann and Nunamaker (1968) have used small (220 pF) memory capacitors that are discharged via field-effect transistors (Fig. 4.8.0.1(c)). In their tests, and wishing to minimize the amount of material in the chamber for reasons of scattering, the experimenters used planes of very fine (0·001 in.) wires. The F.E.T. read-out system worked well, even though a consequence of the fine wires was that the discharge was limited to about 20 nC—about 1% of the charge necessary for core or magnetostrictive read-out.

A large wire chamber spectrometer with a capacitor memory has been reported from SLAC (Coombes *et al.*, 1970). Twenty planes (gaps) having areas ranging from 120 × 120 to 200 × 240 cm² were designed with stretched wires spaced by 1 mm. The read-out was from wires on the low-voltage side only. Each of these wires was connected to a 5 nF capacitor which was normally charged to 15 volts. A spark to the wire discharges the capacitor, and it remains discharged until its associated F.E.T. is strobed and the information read out. The high-voltage electrode was an identical wire plane; its wires were placed parallel to those of the low-voltage electrode, but each wire was connected to the HV bus through a series resistor of 500 Ω, in order to limit the individual spark currents. Although no quantitative tests were made of the efficiency, the practitioners observed hundreds of sparks in single extensive air showers (cosmic rays). They noted that the capacitor read-out appeared to be better than an optical viewing system, for the former would detect sparks that were invisible to the dark-adapted eye.

A capacitor is not the only conceivable memory that may operate in a magnetic field. An alternative possibility (Schüller, 1969) depends on the finite time of "reverse recovery" in a standard diode. If a spark pulse passes through the diode, and if before the diode has recovered (i.e. within about 60 μsec) a pulse of opposite polarity is applied in the same direction, then the latter will pass and can be recorded. No prior spark means no output. At the time of writing, this idea can only be described as tentative—although a proposal does exist to read out 8000 wires in 40 μsec by this means.

4.9 VIDICON OBSERVATION OF SECONDARY SPARKS

Krienen (1970) has demonstrated a method for reading out the information stored in spark chambers that operate in magnetic fields. With conventional crossed-wire planes, he caused the sparking wires to produce secondary sparks at their free ends—which would normally be extended to outside the magnetic field. The secondary sparks could then be recorded with a vidicon camera (see section 4.3.1) and subsequently, digitized.

Referring to the sketch (Fig. 4.9.0.1), the spark chamber proper is on the right (for simplicity, the perpendicular wires are drawn in parallel). The electrodes are coupled in such a fashion that a primary spark causes two projected sparks, one for each of the two participating wires (look to the left). On pulsing, the load presented to the discharging capacitance C_c, is $C_a + C_d + C'_a$ shunted by $R_1 + R_2 + R'_1$. To ensure that most of the voltage occurs across the chamber, and to prevent spurious sparking at C_b, $C_a \gg C_d$. Also $C_c/C_d = 2$ to 4. On sparking, the voltage between the participating wires drops to a low value. This produces a potential difference between the wires and their neighbours, and so an electromagnetic wave

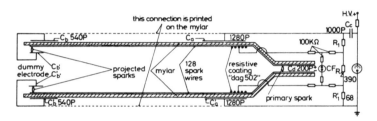

Fig. 4.9.0.1 Sketch of a model 128 crossed-wire chamber with readout via secondary (projected) sparks. (For simplicity, the perpendicular wires are shown parallel.) An event results in a primary spark on the right, and this is followed by secondary sparks on the left, at the ends of the participating wires. (Krienen, 1970)

propagates to the open ends of the wires (printed on mylar, in this case). This causes overvolting across the 1·7 mm gaps formed by the wire ends and the dummy electrodes, and secondary sparks occur.

Krienen found it best to use helium–neon (30/70) plus some organic vapour (13 Torr, 2-propanol) in the 8 mm gap spark chamber; but either pure helium or pure argon in the projection chamber, for the latter have a high threshold and a high luminance. The Mullard Plumbicon camera tube was used. It was hypothesized for this tube that the projected spark points could be arranged in 144 rows of 256, giving a digitizing capacity for one camera of 36,864 wires. Larger systems would require more cameras. It remains to be seen whether this method becomes widely used.

4.10 INDUCTIVE LOCATION OF SINGLE SPARKS

This method, which has been developed by Springer *et al.* (1969), employs the strong magnetic fields that are generated by sparks. To measure the x coordinate of a spark between two plane electrodes, two wire loops S' and S'' are shaped in the fashion shown in Fig. 4.10.0.1. A spark current I_p will induce voltage pulses U'_x and U''_x in the loops according to

$$U = -L \frac{dI_p}{dt}.$$

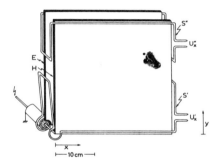

Fig. 4.10.0.1 Inductive location: A spark between electrodes H and E induces voltage pulses in the loops S' and S'' whose amplitudes depend on the x coordinate of the spark. (Springer *et al.*, 1969)

The coefficient of mutual inductance L will depend on the x coordinate of the spark. The signals from the two loops may be added together with a ferrite transformer to give a secondary output of amplitude $U_x = U'_x + U''_x$. To determine L independently of variations in the primary current, dI_p/dt was measured for every spark by observing the output U_p of a small loop wound around the HV input lead. Then we have

$$L(x) = \frac{U_x}{U_p}.$$

The particular form of the loops S' and S'' resulted in $L(x)$ being practically a linear function of x, for all y coordinates.

The read-out may be calibrated accurately, and it was found that a good fit for all calibration lines was given by

$$L(x) = a + bx + cx^2.$$

An identical set of loops may be added to the system to give the y coordinate. The authors quoted a measurement precision of 0·26 mm (FWHM) in a 30 × 30 cm^2 chamber. A drawback is that the system may only handle single sparks, but it has the advantage of being unaffected by magnetic fields.

4.11 ASSORTED METHODS OF RECORDING

4.11.1 The delay-line chamber

In 1962 Charpak suggested that sparks may be located in a spark chamber if one of the electrodes has the structure of a distributed delay-line. In this event, the electrical signal would travel from the spark position to the edges of the electrode with a relatively low velocity. The delay between the arrival of the signals at two opposite edges would indicate the spark position.

In the prototype, the high-voltage electrode was a simple metal plate. The low-voltage electrode, which acted as the delay-line, was constructed from an aluminium plate, covered with a layer of mylar, on top of which a spiral of copper wire was wound. Thus the spark crossed between the HV plate and one of the turns of the winding. The signal delay is $\sqrt{(LC)}$ per cm and the line impedance is $\sqrt{(L/C)}$, where the inductance L and the capacitance C per unit length will be determined by the dimensions. At the edges the signals are observed across terminating resistors. Experience with several prototypes, the fastest having a delay of 200 ns in 40 cm, led Charpak to estimate that an accuracy of 1 mm is feasible.

4.11.2 Digitizing with photodiodes

This neat method, which was invented by Diambrini and Giannini (1966), depends upon a specially designed optical system. In Fig. 4.11.2.1, the cylindrical lens L_1 focuses a sharp line-image of the spark on the screen S. This screen is divided into eight horizontal bands which are divided into opaque and

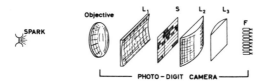

Fig. 4.11.2.1 Digitizing camera: a vertical line-image of a spark is produced on the mosaic screen S, and selected pencils of light strike appropriate photodiodes. (Diambrini and Giannini, 1966)

transparent regions according to a binary logic. Behind each band is a photodiode. The screen S is at the focus of the lens combination L_2L_3 to ensure that light passing through a particular band lies in a pencil aimed at the corresponding photodiode.

Thus, with this camera, the light from a spark is used to define its location with an eight-digit binary figure. Although the choice of mosaic logic blocking in the screen S is fairly wide, it is best to choose a "reflected binary" code (Fig. 4.11.2.2).

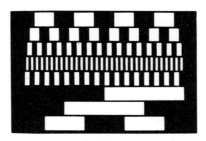

Fig. 4.11.2.2 The "reflected binary" pattern of the mosaic screen in the digitizing camera. (Diambrini and Giannini, 1966)

With this logic, the binary numbers increase regularly as the position of the spark is moved horizontally. This reduces the possibility of error when sparks lie on the boundary between two zones.

In the prototype an accuracy of ± 0.5 mm was obtained, but this would improve with further development. As the authors point out, the method has the advantage of being relatively simple, but the disadvantage of being able to record only single sparks.

4.11.3 Direct recording with magnetic tape

When a spark crosses between two electrodes it causes a magnetic disturbance at the electrodes. Quercigh (1964) devised a chamber consisting of two stainless-steel mesh electrodes. Just beneath one, he ran a magnetic tape. A spark to a particular point of the mesh caused a magnetized spot in the tape. The tape was subsequently scanned with a standard recorder unit. Quercigh estimated the attainable precision to be 1 mm, and, of course, multiple simultaneous sparks could be recorded.

CHAPTER 5

Wide-gap Spark Chambers

5.1 INTRODUCTION

IN the wide-gap spark chamber, a bright spark actually follows the line of a particle's path. This phenomenon is a consequence of neighbouring avalanches, created along the ionized trail, merging together to form a plasma in which the spark current may pass. This merging can be reliable if the particle's direction lies within about 30° of the applied field. This property of track delineation is to be contrasted with the usual narrow-gap case in which the spark develops at right angles to the electrodes.

A conventional counter arrangement may be used to initiate the transfer of a high-voltage pulse to the central electrode of a double-gap chamber (Fig. 5.1.1.1). A field of the order of 10 kV/cm is necessary, so that with an electrode spacing of 20 cm, a pulsed voltage of 200 kV is required. This is normally provided by a Marx generator. In the usual neon/helium mixture, well-defined sparks will indicate the particle's orbit (Fig. 5.1.1.2).

Following early attempts by Charpak (1957), Alikhanian (1963) was one of the first to develop a useful wide-gap chamber. At the time the incentives were clear. On the one hand, it was not always easy to define a particle's trajectory exactly from a row of discrete sparks; and on the other, the multiple scattering of a beam in the numerous electrodes of a multi-gap chamber

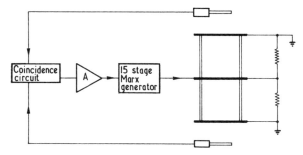

Fig. 5.1.1.1 Elements of a wide-gap spark-chamber system.

172 WIDE-GAP SPARK CHAMBERS

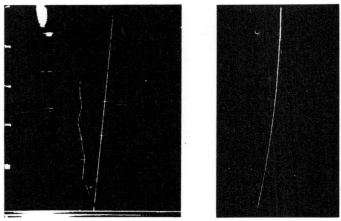

Fig. 5.1.1.2 Tracks in a 40 cm gap spark chamber. (Garron *et al.*, 1965)

was a nuisance. In recent years the development of low-density wire-plane chambers has reduced the problem of multiple scattering, and their capacity for automatic digitization has made them popular. These factors, and the advent of streamer chambers which show a better angular response, have meant that the wide-gap spark chamber has not become a widely used tool in particle physics. Nevertheless, in some circumstances, they can be very useful.

5.2 CREATION OF THE SPARK CHANNEL

In designing wide-gap spark chambers it would be advantageous to know exactly how a spark channel is created along an ionized path at an angle to the applied field. Photographs of merged streamer tracks (see Fig. 6.1.4.1) indicate that the luminous filaments represent a faithful record of the tracks—in spite of the ripples. It is thought that the avalanches will tend to follow a field that is the resultant of the applied field and the space charge field of the neighbouring avalanche (Fig. 5.2.1.1). It is easy to imagine that electrons in the head of one avalanche will be attracted towards the positive tail of its neighbour.

Evans (1969) considered the growth of equally spaced avalanches along an inclined track. He calculated the changes

CREATION OF THE SPARK CHANNEL

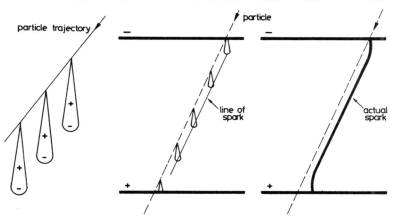

Fig. 5.2.1.1 Illustrating the alignment of a spark along a particle trajectory due to a combination of space-charge field and applied field.

in charge density and was able to plot the resultant field at various times. Figure 5.2.1.2 shows that after a certain time the field lines orient themselves along the track direction and the avalanches merge to form a continuous tube.

The calculations refer to the ideal case of equally spaced avalanches all developing uniformly. In practice there are fluctuations in the spacing of the sites of primary ionization and these are likely to be a cause of the irregularities (ripples) that one observes (Fig. 6.1.4.1). In addition, these irregularities will be exacerbated by the fluctuations in the rates of growth of individual avalanches. At large angles the effect will be most noticeable; when an avalanche happens to develop in advance of its neighbours, the space-charge field may not be enough to constrain the discharge to the particle's track. Rather, the avalanche may propagate a streamer towards the electrodes, and thus bring about a discontinuity in the filament (cf. Fig. 5.5.2.2).

The experiments of Caris *et al.* (1968) show that in the later stages of development the discharge grows at a "faster-than-exponential" rate (Fig. 5.2.1.3), i.e. faster than the rate predicted by the Evans model. At high-current densities it is plausible that two-stage ionization processes become important. Examples would be ionization of an excited atom by a low-energy electron, or by a photon, or by collision with another excited atom. After

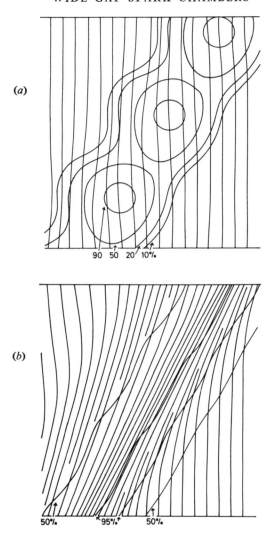

Fig. 5.2.1.2 The merging of equally spaced avalanches along an inclined track. Calculated contours of electron density are shown, together with the directions of the electric field. (*a*) refers to a time 82·8 nsec, and (*b*) to 97·9 nsec, from the start of the avalanches. The parameters are: neon; 690 Torr; 6·73 kV/cm; track angle 30°; avalanche separation 2 mm; and the anode is at the top of each figure. (Evans, 1969)

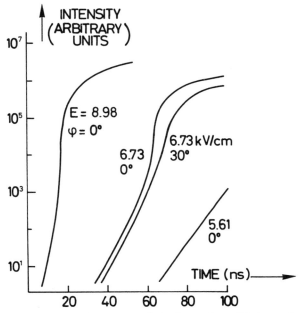

Fig. 5.2.1.3 Temporal growth of the intensity of light emitted from a spark chamber for various values of electric field E and angle ϕ between the direction of the particle beam and the electric field (689 Torr). (Caris et al., 1968)

considering these possibilities, however, Evans concluded that according to his model the theoretical value for the ionization would still be too low by a factor of 1000 or so. Also the fact that sparks are normally seen to be only 1 or 2 mm wide, suggests that photoionization of ground-state atoms just outside the track is not significant.

As an alternative explanation for the rapid build up of current, Evans points to the statistical fluctuations within a spark. When the density has become about 1 A/cm^2, the spark becomes a low-resistance conducting channel joining the plates. Due to fluctuations extending over many avalanches, the resistance will not be uniform and so a large-scale redistribution of voltage occurs. In the regions of higher field, ionization will proceed at a higher rate. In turn this will increase the field in other regions, and so on. Thus, because $\partial^2 \alpha/\partial E^2$ is positive here, these fluctuations might explain the rise in the experimental curves.

A spark will not indicate the actual path of an inclined trajectory. Rather, it follows a parallel line that is displaced by a distance related to the drift of the electrons in the initial avalanches (Fig. 5.2.1.1). This distance depends on the rise time of the applied field. The effect leaves a hiatus at the cathode and must be responsible for the distortion of the track in the field direction that is observed at this electrode (Peter *et al.*, 1963). A similar distortion is seen at the anode, but this is more difficult to explain. No doubt it is due to local space-charge field conditions. Charpak *et al.* (1965) speculated that electrons might be extracted from the anode. In practice, the distortions become less significant, and indeed the ability of the spark to register large inclinations improves, for larger electrode separations and higher pressures. One way to correct the parallel displacement of the track is to apply a small d.c. voltage in such a manner that the electrons are drifted in the opposite direction during the delay before the high-voltage pulse is applied. Garron *et al.* (1965), for example, found that the drift effects of a 22 kV/cm field would be negated in the 500 nsec delay, if a steady 100 V/cm field existed.

5.3 HIGH-VOLTAGE PULSE GENERATION

5.3.1 The Marx generator

The success of wide-gap chambers has depended upon the production of voltage pulses with rise-times measured in nanoseconds and amplitudes in hundreds of kilovolts. Almost without exception, the Marx (1924) generator has been used. Its operation is simple: a row of low-inductance condensers (C_0) is charged up slowly in parallel and then, when required, discharged rapidly in series. The low-impedance connections are spark gaps that fire almost simultaneously. The voltages add and the output pulse is the sum. Figure 5.3.1.1 is an example. A moderate d.c. voltage (30–60 kV) is applied to the capacitors through a number of high resistances. These resistances appear essentially infinite during the pulsing and may be ignored. Thus with a 40 kV supply, five stages would give 200 kV.

The Marx is normally triggered by applying a pulse to a

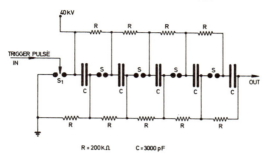

Fig. 5.3.1.1 The elements of a Marx generator.

third electrode in the first gap. This gap breaks down. If there are n stages, each charged to voltage U_0, this first breakdown will result in an over-voltage of $U_0/(n-1)$ on each of the remaining spark gaps. This is because a step voltage of U_0 is applied to the $n-1$ gap capacitances (C) in series. The spacings of these remaining gaps may be chosen to breakdown with this over-voltage.

This process is rather slow because the degree of over-voltage is small. A much faster breakdown of the Marx is achieved with the use of small coupling condensers (C_c) connected as in Fig. 5.3.1.2. With this arrangement (if $C_0 \gg C_c \gg C_s$), when the first gap fires, the second becomes over-voltaged by U_0, and thus rapidly breaks down. As a consequence, the third gap is over voltaged by $2U_0$, and so on. When all the gaps have fired in quick succession, a fast-rising output pulse emerges.

Gygi and Schneider (1964) have analysed the behaviour of a Marx in conjunction with a chamber. In Fig. 5.3.1.2(*a*), one expects an output voltage of

$$\bar{U} \approx nU_0 \frac{C_0}{C_0 + 2nC_c}$$

to develop across the Marx capacitance $C_1 = C_0/n$. Assuming that the switching time of n spark gaps is faster than the transient times of the circuit, they can be represented by one switch S_1. Figure 5.3.1.2(*b*) shows the equivalent circuit, with L_1 representing the combined inductances of the sparks and condensers in the Marx; C_2 the capacity of the chamber; and Z the terminating resistor or characteristic impedance of the line.

The Laplace transform for the voltage over Z if the switch is closed at $t = 0$ is given by:

$$u = \frac{\bar{U}}{L_1 C_2} \frac{1}{s^3 + \frac{1}{C_2 Z} s^2 + \frac{C_1 + C_2}{L_1 C_1 C_2} s + \frac{1}{L_1 C_1 C_2 Z}}.$$

If the roots of the denominator are:

$$s_1 = a; \quad s_2 = \alpha + i\beta; \quad s_3 = \alpha - i\beta$$

the solution is:

$$\frac{U}{\bar{U}} = \frac{1}{L_1 C_2 (a^2 - 2\alpha a + \alpha^2 + \beta^2)}$$

$$\left[e^{at} - e^{\alpha t} \left(\frac{a - \alpha}{\beta} \sin \beta t + \cos \beta t \right) \right]. \quad (1)$$

Figure 5.3.1.2(c) shows the output voltage for particular values of L_1, C_1, C_2 and Z.

For many purposes it is useful to shorten the pulse by clipping its tail with a shunting-spark gap S_2, in which case the equivalent circuit becomes (d). To calculate the waveform one assumes that S_1 has already been closed, and that at time $t_1 = 0$, S_2 is closed. The initial conditions ($\bar{U}_1$, $\bar{U}_2$, J_1) can be worked out from Eq. (1), and if for simplicity one takes $C_2 = 0$, the Laplace transform for the circuit becomes

$$u = \frac{Z}{L_1} \frac{\bar{U}_1 s - L_1 J_1 s^2}{s^3 + \frac{L_1 + L_2}{L_1 L_2} Z s^2 + \frac{1}{L_1 C_1} s + \frac{Z}{L_1 L_2 C_1}}.$$

If the roots of the denominator are:

$$s_1 = a; \quad s_2 = \alpha + i\beta; \quad s_3 = \alpha - i\beta;$$

the solution is:

$$\frac{U}{\bar{U}} = \frac{Z}{L_1 \beta (a^2 + \alpha^2 + \beta^2 - 2a\alpha)} \left[a\beta \left(\frac{\bar{U}_1}{\bar{U}} - a \frac{L_1 J_1}{\bar{U}} \right) e^a \right.$$

$$+ \left\{ \left[(a\alpha^2 - \alpha^3 - \alpha\beta^2 - a\beta^2) \frac{L_1 J_1}{\bar{U}} + \frac{\bar{U}_1}{\bar{U}} (\alpha^2 + \beta^2 - a\alpha) \right] \sin \beta t \right.$$

$$\left. - \left[(\beta^3 + \alpha^2 \beta - 2a\alpha\beta) \frac{L_1 J_1}{\bar{U}} + \beta a \frac{\bar{U}_1}{\bar{U}} \right] \cos \beta t \right\} e^{\alpha t} \right].$$

Figure 5.3.1.2(e) shows the waveform with the condition that S_2 was closed at the moment of maximum amplitude for the $C_2 = 0$ curve in (c). The calculations suggest a possible rise time of 2 ns.

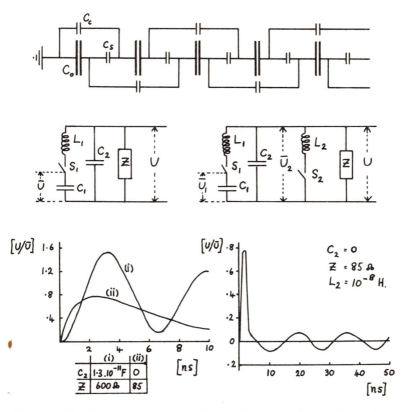

Fig. 5.3.1.2 Marx generator with coupling capacitors. Equivalent circuits and waveforms for particular values. In all curves, $L_1 = 10^{-7}$H and $C_1 = 7.10^{-11}$F (see text). (Gygi and Schneider, 1964)

Gygi and Schneider (1964) constructed an elegant ten-stage Marx along these lines. The Marx had an overall delay of 10 ns and produced a pulse of 200 kV with a rise time of 2 ns. Although these extraordinarily good figures are not necessary for many purposes, it is valuable to consider the design. The low-inductance storage capacitors were cylinders (25 mm diameter, 7·5 mm height) of barium titanate that had an effective capacity of 700 pF at 100 MHz when stressed with 20 kV. At their ends were thick disc electrodes of stainless steel, whose rounded edges formed 1 mm spark gaps. The condensers were arranged in two adjacent stacks with the barium titanate alternating with polystyrene—the latter constituting the coupling capacitors. The cylinders were covered with teflon envelopes, and when the whole device was enclosed in 7 atmospheres of nitrogen, a voltage of 22 kV could be withstood indefinitely.

The rapid breakdown of a spark gap is facilitated if there are free electrons in the gap prior to the signal. A constant corona discharge from an adjacent pin is one possible source of continuous supply: the positive ions created at the point of the pin, diffuse towards the cathode and on impact liberate electrons. These then move across the gap and are absorbed by the anode. Gygi and Schneider used tungsten or platinum pins, radius 1/100 mm, with the current limited to 1 μA. They found that a high efficiency of electron emission was obtained by adding a little water vapour—effectively to lower the work function of the metal surface.

Many other Marx have been reported. Garron et al. (1965) constructed a 15 × 30 kV Marx with a delay of 100 nsec. To obviate the variable breakdown which accompanies atmospheric changes of pressure and humidity, the whole system was enclosed in an airtight box. Rapid discharging was aided by allowing the gaps to "see" each other, i.e. ultra-violet from the first breakdown liberates electrons in the remaining gaps. Keller and Walschon (1966) produced a five-stage, 22 kV/stage coaxial design with a delay of 20 nsec. The whole unit was pressurized with nitrogen at 1·8 atmospheres, and the interior of the pressure vessel was painted white to provide an optical linkage between the 2 mm gaps. The author and S. R. Mishra (1969) employed a four-stage, 60 kV/stage, coaxial design that operated in air. A large Marx has been built at SLAC (Bulos et al., 1967) (Fig. 5.3.1.3). By using both positive and negative polarities for d.c. charging, the thirty-four stages only needed seventeen spark gaps. The generator is capable of running with a charging voltage of ± 50 kV, producing an output pulse of

THE MARX GENERATOR

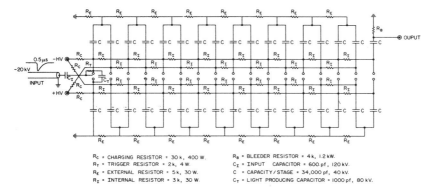

R_C = CHARGING RESISTOR = 30 k, 400 W.
R_T = TRIGGER RESISTOR = 2 k, 4 W.
R_E = EXTERNAL RESISTOR = 5 k, 30 W.
R_I = INTERNAL RESISTOR = 3 k, 30 W.
R_B = BLEEDER RESISTOR = 4 k, 1.2 kW.
C_I = INPUT CAPACITOR = 600 pf, 120 kV.
C = CAPACITY/STAGE = 34,000 pf, 40 kV.
C_T = LIGHT PRODUCING CAPACITOR = 1000 pf, 80 kV.

Fig. 5.3.1.3 Marx generator with 17 gaps and 34 stages, to drive the SLAC 2 metre streamer chamber. (Bulos *et al.*, 1967)

1300 kV! However, normally it is run at ±20 kV, to produce 600 kV for the streamer chamber. In the DESY fourteen-stage 500 kV generator (Fig. 5.3.1.4) the spark gaps contained pressurized nitrogen and the resistors and condensers were immersed in oil (Ladage, 1969).

Fig. 5.3.1.4 Marx generator at DESY. (Ladage, 1969)

5.3.2 Miscellaneous HV generators

The Marx generator has been universally popular as the source of high voltage for wide gap and streamer chambers. However, several alternatives exist which are capable of creating voltages in the range of 10^5–10^7 volts and it is possible they may yet play a role. We shall mention a few examples; detailed information and their practical performance may be obtained from the original articles by Fitch and Howell (1964) and Martin and Smith (1967).

Referring to Fig. 5.3.2.1, in contrast to the Marx principle, (*a*) illustrates the production of high voltages by inversion of alternate E-vectors. Such an inversion may be achieved with an electromagnetic wave in a transmission line; e.g. the closure of a switch at one end of a charged line leads to an inversion of the E-vector at the open end that persists for twice the propagation time of the line. (*c*) shows a stack of charged strip lines where alternate lines may be switched simultaneously. The output pulse $V_0 = 2$ nV (i.e. static + incident + reflected voltages). In practice, resistive losses, capacitive coupling and switch impedances cause some attenuation and loss of rise-time. The complication of multiple switching can be avoided with the circuit (*d*). Coaxial lines may be used in this fashion but the amplification can be curtailed as a result of the coaxial lines having, in general, higher impedances than strip lines.

The LC generator (*b*) also employs vector inversion. In the lumped circuit ($\tau = \pi(LC)^{\frac{1}{2}}$) the output voltage is

$$V_0(t) = nV[1 - \exp(-\alpha t) \cos \omega t]$$

where $\omega^2 = 1/LC$, $\alpha = R/2L$ and R is the resistance. The peak voltage is

$$V_0 = nV\left[1 + \exp\left(\frac{-\pi R}{R_c}\right)\right]$$

where $R_c = 2(L/C)^{\frac{1}{2}}$. In comparison with the Marx, the output waveform is quite different, the number of switches is halved, and the output impedance can be much lower because it does not include the internal switches.

A spiral generator is shown in (*e*). It is really a twin spiral line. A switch at the mid-point initiates waves in one line only. The

MISCELLANEOUS HV GENERATORS 183

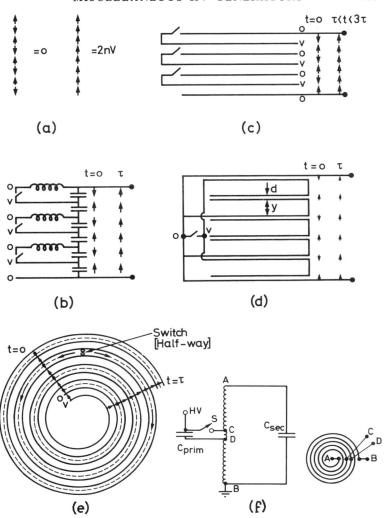

Fig. 5.3.2.1 Miscellaneous HV generators (see text).

waves convert electrostatic to electromagnetic energy in the active line leaving the electric vectors in the passive line unopposed, so that the voltage between the inner and outer turns builds up to nV by the time the waves reach the unterminated ends of the line. The reflected waves convert the electro-magnetic energy back to electrostatic energy of the opposite sign, and the output voltage eventually rises to 2 nV. Further reflexions at the switch, etc., reverse the whole process, thus giving the pulse a triangular shape. Fitch and Howell show a 750 kV spark produced by a compact 20 kV spiral generator.

Finally, there is the autotransformer (f) in which a high-voltage capacitor is suddenly switched across a single-turn primary with a spark gap—and the high-voltage appears across the secondary. Pulses greater than 5 MV have been produced in this way. Martin and Smith describe a remarkably simple autotransformer—just sixteen turns of copper foil with some mylar insulation, all immersed in copper-sulphate solution—and lo and behold, 50 kV begets 800 kV!

5.4 EXAMPLES OF WIDE-GAP SPARK CHAMBERS

5.4.1 Harvard chamber in a magnetic field

A good example of a wide-gap spark chamber has been described by Garron et al. (1965). A fifteen-stage Marx generator operated at 30 kV transfers a 50 nsec, 450 kV pulse to the central electrode of a chamber with two 40 cm gaps. Vacuum-tight glued perspex walls are sealed with "O" rings to the aluminium plate electrodes. With 90% neon/10% helium, the tracks in Fig. 5.1.1.2 were obtained. Pure neon gave similar tracks, helium less stable. For single tracks, the spark intensity slowly decreases as the inclination angle increases up to 40°, beyond which it decreases very rapidly. When many tracks were present (up to twenty had been seen) some were often too faint to be recorded. Simultaneous exposures at $f/45$ and $f/16$ were recommended for multiple particle events. Sensitive times were typically 5 μsec.

Attention was paid to distortions that might arise if the electric field were not uniform. After tests on tracks near the side walls of the chamber, Garron et al. concluded that two approaches gave an acceptable performance: one could either extend the electrodes 30 cm beyond the sensitive volume, or one could install a multiplate shielding condenser against the perspex walls.

In a magnetic field one expects an $E \times B$ displacement of the tracks. To detect the effect of this the chamber was placed in a

13 kGauss magnet. A shielding condenser was used on the side near the single solid pole face, and 24 cm from it. Photographs were taken through the 1·4 m diameter hole in the second pole. By comparing the curvature of tracks of cosmic rays (>1 GeV/c) in the two chambers, it was estimated that the error in measuring the curvature amounted to $\pm 0{\cdot}55 \times 10^{-4}$ cm^{-1}. This corresponds to a momentum determination of $\pm 1{\cdot}4\%$ on a 1 GeV/c particle.

Also at Harvard, Knasel et al. (1965) showed how a chamber may be used when an intense beam of charged particles is passing through it. Normally, in neon, the whole available high-voltage energy would discharge in the beam trail and one would find a zero efficiency for a single out-of-beam particle. However, if the beam region is filled with helium enclosed in a transparent thin mylar bag, a single particle will be detected in the surrounding neon. And photography through the bag is possible. To assess the efficiency, an off-beam particle had to trigger a three-counter telescope with a 5 nsec coincidence resolving time. With a 60–70 kV high-voltage pulse and a 76 mm chamber gap, when a beam of $2{\cdot}4 \times 10^{11}$ protons per second passed through the helium, the efficiency remained 100% in the 89% neon, 10% helium, 1% argon surrounding gas. The sensitive time was 9 μsec which meant that $2{\cdot}2 \times 10^6$ beam particles had passed in the sensitive time—and yet had no effect.

5.4.2 Digitized chamber at CERN

Track following wire-plane chambers have been successfully employed in the CERN boson spectrometer. (Chikovani, Laverrière and Schubelin, 1967.) The wide (10 cm) gap chambers combine the advantages of good angular resolution, high multi-track efficiency, the absence of spurious tracks or breakdowns and automatic magnetostrictive read-out.

The chamber construction is shown in Fig. 5.4.2.1. Four wire planes, alternatively crossed, define two gaps. They are composed of parallel 0·1 mm diameter phosphor bronze wires, spaced at 1 mm, and wound on 1 cm thick fibre-glass frames. The size (160 × 160 cm) requires steel supports which are araldited to the fibre-glass. Plexiglass walls separate the central high-voltage electrodes from the outer earthed planes. Rubber gaskets allow demounting and the whole chamber is made gastight by the thin mylar windows enveloping the earthed planes. The wires are free at one end, and soldered to a copper bar at the other for the electrical connections. The beam region is deadened by inserting a 6 cm diameter, 3 cm thick styrofoam disc held with nylon strings.

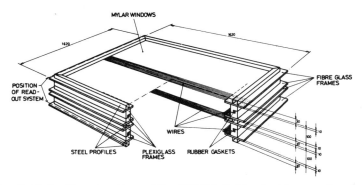

Fig. 5.4.2.1 Construction of the CERN digitized wide-gap chamber. (Chikovani *et al.*, 1967)

The 100–150 kV pulse was derived from a seven-stage Marx generator with an equivalent capacity of 860 pF. The pulse rose in 25 nsec and the length was adjustable between 50–100 nsec with a shunting spark gap. The Marx repetition cycle was reduced to 5 msec by charging the capacitors in parallel from a buffer 14 μF capacitor across the power supply. The memory time in straight 70% neon/30% helium was 10 μsec, but the addition of 5% argon and small quantities of quenching freon-12 enabled the times to be reduced to about one microsecond.

The technique of magnetostrictive read-out has been described in section 4.5. In the present case the fact that four scalers were attached to each pick-up coil meant that up to eight sparks could be registered. Figure 5.4.2.2 shows a photograph of sparks occurring in the chambers.

Fig. 5.4.2.2 Simultaneous particle tracks in the CERN wide-gap spark chamber. (*Photo CERN*)

5.5 PERFORMANCE

5.5.1 Response to inclined tracks

In their study of chamber characteristics, Aronson *et al.* (1967), at Princeton, measured the efficiency for single particles as a function of angle. Their 41 cm gap chamber is shown in Fig. 5.5.1.1. The electrodes were 0·001-in. stretched aluminium foils. Correction loops of fine wire, interconnected with resistors, and wound round the 1 in. perspex walls of the chamber, ensured a uniform field. With narrow-gap chambers mounted on either side to define a track, the efficiencies, when the chamber was rotated in a 395 MeV/c beam, were measured to be 99·8, 99·6, 99·1 and 97·3% at inclinations of 0, 10, 20 and 30° respectively. However, there is nothing fundamental about those percentages: they just reflect operating experience at Princeton. In principle, one would expect 100% efficiency for single particles at small angles in a wide-gap chamber, as there are plenty of ionization electrons along the track to initiate the discharge.

At Argonne, Keller *et al.* (1966) found their efficiency to be

Fig. 5.5.1.1 The Princeton wide-gap spark chamber. (Aronson et al., 1967)

essentially 100% for all angles up to 45°. However, in a thorough study with a double 10 cm gap chamber, they observed that the brightness of the tracks declined rapidly with inclination (Fig. 5.5.1.2). An efficiency below 100% is therefore likely to concern the high-voltage pulsing or the photography.

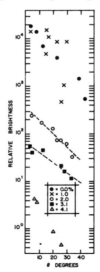

Fig. 5.5.1.2 Relative brightness of sparks as a function of angle between track and electric field. (Dashed lines indicate concentrations considered optimum – see Fig. 5.5.2.1.) (Keller et al., 1966)

5.5.2 On gases and track quality

The Argonne studies also included the effects of adding alcohol vapour to the chamber gas (90% neon/10% helium). Figure 5.5.2.1 shows how the quality of the tracks improves with increasing concentrations. The width decreases and the ripples diminish. The spread of track centres (σ) about a fitted straight line is reduced in the fashion 0·14, 0·11, 0·08, 0·075 and 0·07 mm for concentrations of 0, 1, 2, 3, 4% respectively. But the

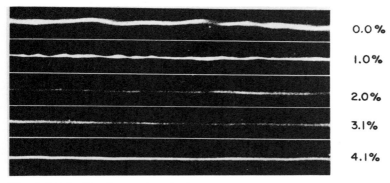

Fig. 5.5.2.1 Portions of typical spark-chamber tracks with various percentages of ethyl alcohol added to Ne-He. Length of track portions: 0·0%, 2·6 in.; other 3·2 in. (Keller, *et al.*, 1966)

luminosity of the tracks goes down too. The addition of benzene (2·5%) and ethylene dichloride (2%) had like consequences. Keller *et al.* attributed this behaviour to the vapour's capacity for absorbing ultra-violet.

Charpak *et al.* (1965) suggested that iodine vapour would play a similar role; that its wide band absorption in the ultra-violet would prevent avalanches diverging away from the track. They pointed out that the Penning effect would also enhance the multiplication. The excitation energy of the metastable states from He (19·8 eV), Ne (16·53 eV, 16·62 eV) or A (11·49, 11·66 eV) can be transferred to an electron from the I_2 molecule (9·7 eV ionization potential).

Kotenko *et al.* (1967) in Moscow have reported analogous results with ethanol and with water vapour. After a series of measurements on spark tracks that had been initiated with various delays (2·8, 12, 30, 50 μsec), they concluded that the main action of the vapour molecules was loosely to attach electrons near the sites of primary ionization. In other words, the free diffusion of the electrons was prevented. But the coupling energy is so small that the electrons are easily released when the strong field is applied. The effective diffusion coefficients (D) were measured to be 78, 7·9, 0·34, 0·29 cm^2/sec for high-purity neon, ordinary neon, neon plus saturated water vapour, and neon plus saturated C_2H_5OH vapour respectively.

Assuming an electron capture probability by water molecules at 20°C, $h = 4 \times 10^{-4}$ and a collision cross-section of 25×10^{-16} cm^2, Kotenko et al. calculated that the number of free electrons in saturated water vapour would diminish by a factor 10^3 in 1 μsec. They suggested that the capture of an electron leads to the creation of complexes like $(2H_2O)^-$. Such a process is possible in vapours which possess large dielectric constant (ε) and a considerable dipole moment. The same applies for ethanol. Figure 5.5.2.2 indicates the practical benefit of adding C_2H_5OH.

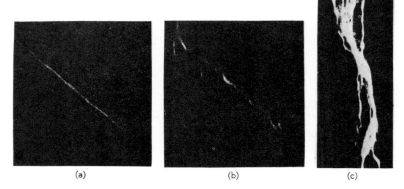

Fig. 5.5.2.2 Inclined tracks in a wide-gap chamber filled with (a) Ne + C_2H_5OH; (b) Ne; (c) neon of high purity ($T_d = 2·8$ ns). (Kotenko et al., 1967)

5.5.3 Recording simultaneous tracks

In a narrow-gap chamber the sparks often grow from a single avalanche. When many simultaneous trajectories are present, the considerable fluctuations in the time development of the corresponding avalanches may lead to all the available current passing through the first-created spark channel. When this happens, the other sparks are "robbed" and are not recorded.

In the wide-gap chamber, a track spark develops from, say, ten or more avalanches. One would therefore not expect significant statistical fluctuations in the development of separate "simultaneous" spark channels. Bolotov and Devishev (1964) used three chambers (10 cm gap, neon, aluminium electrodes) to measure the efficiency for detection of showers. They found that the chance of observing a track in the middle chamber, corresponding to sparks in the outer chambers, lay between 98 and 100% for showers containing up to fifty particles. The detection was good for angles up to 40°. In practice, it must be ensured that the photography is capable of registering even the weak trails which are really only merged streamers (see section 6.1.4) (Fig. 5.5.3.1).

Fig. 5.5.3.1 Part of an extensive air shower registered in two spark chambers (10 cm gaps). (Bolotov and Devishev, 1964)

An interesting phenomenon is observed when neutral particles decay within the sensitive volume of a spark chamber. Catura and Chen (1965) found that in addition to the well-delineated prongs of the two charged particles, a diffuse discharge also occurred from the vertex towards the opposite electrode (Fig. 5.5.3.2). This discharge must carry the current in the absence of an ionized trail.

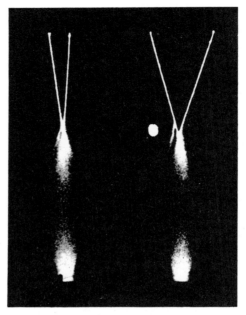

Fig. 5.5.3.2 A double prong effect seen with a 40 cm wide-gap spark chamber. (Catura and Chen, 1965)

5.5.4 Memory and recovery times

When a spark-chamber system is triggered the aim is to display only the selected events. Tracks due to particles passing a few microseconds before are an embarrassment. One would like a short sensitive (memory) time so that the chamber can be operated in high-intensity beams. In narrow-gap chambers, a constant low-voltage field ensures that ionization electrons are drifted away swiftly in the absence of a trigger. With a 30 cm gap chamber, if one were to connect 3000 volts across the plates ($E/p = 0.13$ volts/cm/Torr) one sees from Fig. 1.4.2.1 that the electron drift velocity in neon would be 4×10^5 cm/sec. The time to clear the electrons from the gap would thus be 75 μsec—much too long for modern experimentation. And the drift velocities cannot be increased enough by raising the d.c. field.

In pure noble gases (Ne, He, Xe), electrons may remain free for tens of microseconds. However, the addition of small quantities of electronegative gases ensures that the electrons are quickly attached to form negative ions. The effect on the recording efficiency of small concentrations of CH_4 and CCl_4 is shown in Fig. 5.5.4.1. The curves indicate that memory times of about 1 μsec are feasible while still retaining 100% efficiency.

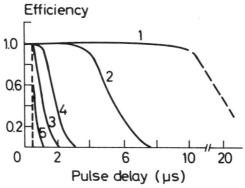

Fig. 5.5.4.1 Particle efficiencies as a function of pulse delay in a wide-gap chamber – at 4·6 kV/cm with 99·6% Ne, 0·4% Ar plus (1) nothing, (2) "nothing" after two year's operation, (3) 0·1% CH_4, (4) 3.10^{-6}% CCl_4, (5) 3.10^{-5}% CCl_4. (Gromova et al., 1965)

When memory times are reduced to this extent chemically, there is a danger that during the delay before the application of the high-voltage pulse, many of the primary electrons will be attached. In this event, the quality of the track might be seriously impaired. In contrast to the results of Gromova et al., Eisenstein et al. (1966) found in a pure neon/helium (90/10) mixture that the effective sensitive times could be as slow as 500 ns. Their sparking conditions were such that only tracks within 500 ns resulted in bright sparks—earlier events (~2 μsec) gave either low-intensity diffuse sparks or were invisible (5 μsec).

Keller et al. (1966) have pointed out that when a d.c. field is applied to compensate for the avalanche drifting and consequent displacement of the sparks, early tracks are drifted further than late ones. Therefore in a double-gap chamber with opposed d.c. fields, the relative displacement of the two sparks indicates the moment of passage. They suggest that this moment can be ascertained to within ±10 nsec by this means.

The aftermath of a spark is a trail of plasma. If the chamber be activated again after only a tiny interval this trail could re-ignite—it might reappear as a track. Clearly, a short recovery time is desirable. With a 90/10 mixture of neon and helium, Keller et al. concluded that in their 10 cm gaps the recovery time was less than 30 msec. On the other hand, when 3·1 % ethyl alcohol was added, earlier sparks often re-ignited, for times up to 100 msec. The re-ignited sparks had a diffuse appearance. For times of this magnitude, a clearing field is useful; with a d.c. field of 100 V/cm, Keller et al. found the recovery time was reduced to about 40 msec when alcohol was present. In the CERN 10 cm chambers, with a small quantity of freon ($< 10^{-4}$) present and a voltage applied, the recycling time could be reduced to 10 msec (Klanner, 1971).

5.5.5 Precision of measurement

Keller et al. (1966) mounted their double 10 cm gap chamber between two banks of six narrow-gap chambers which were used to complete and define tracks of particles passing through the system. In the film, pairs of coordinates were determined on each wide track with a digitizing microscope, and straight lines fitted with least squares. The opposite parallel displacements in the two gaps due to electron drift were subtracted from each other. An analysis gave the following uncertainties in the measurement of a trajectory in the double-gap chamber:

Projected angle of inclination	Uncertainty in angle (σ_θ) (mrad)	Uncertainty in co-ordinate in plane of plates (σ_x) (mm)
0–10°	0·84	0·24
30–40°	1·38	0·31

5.5.6 Ionization effects

A Moscow group (Lyubimov et al., 1964) have studied the connection between initial track ionization and the luminosity and appearance of the sparks. With two 15 cm gap chambers filled with neon (540 Torr) plus 0·03 % propane, they observed that minimum ionizing pions leave fainter trails than protons with 3·5 times the minimum. The histogram shows how, in a series of measurements, the two categories of particles distribute themselves according to photometrically determined luminosity. A factor of ~ 3 corresponds to the ratio 3·5. The authors go on to claim that the proportionality of gas amplification holds up to 10^{10} in a spark discharge (cf. 10^3–10^4 in a proportional counter).

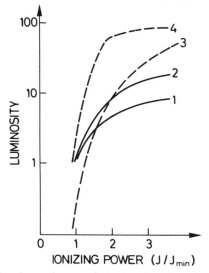

Fig. 5.5.6.1 The dependence of the luminosity of 15 cm sparks on the ionizing power of particles (1, 2 are for neon at 760,500 Torr and fields of 16·8, 24·6 V/cm Torr; 3, 4 are for helium at 460 Torr with 26·8, 27·8 V/cm Torr). (Galaktionov et al., 1965)

In a further study (Galaktionov et al., 1965) the curves of Fig. 5.5.6.1 were obtained. An analysis of the measured fluctuations of luminosity for monochromatic particles was in fair agreement with the Landau distribution.

With quenching propane present, the sparks display "bunches" along their path. These too can be used to estimate ionizing power. In this case the ratio 3·5 was matched by 1·9 in the density of bunches.

In spite of these results it must be stressed that the art is a difficult one. It is not an easy matter to keep the pulse heights and lengths sufficiently constant. The variations of intensity with angle of track inclination would have to be accounted for, and biases in the photography avoided.

CHAPTER 6

Streamer Chambers

6.1 INTRODUCTION

6.1.1 Preliminary

If the primary ion pairs produced by traversing particles could be recorded directly, spark and streamer chambers would be redundant. Of course, they cannot; but the question arises as to what is the minimum amount of gas multiplication that will enable a satisfactory picture of the track to be photographed? It appears that the full development of a conducting spark channel is unnecessary; individual streamers themselves emit sufficient light to register their existence on fast film. This has been the basis of the streamer chamber. By arresting the discharge (see section 1.4.4) at an early stage, a particle trajectory may be indicated by a luminous trail in a wide-gap chamber.

This limitation of the discharge within the gas, and its consequent detachment from the electrodes, has endowed the streamer chamber with the important property that it may manifest particle orbits regardless of their direction, i.e. its response may be isotropic—unlike spark chambers.

In the streamer chamber, the particles need not pass through, or near, the electrodes. By applying big pulses (500 kV, say) to broad electrodes it is possible to sensitize a large volume of gas. The particle's path, and any interactions with the gas, may then be followed for distances of several metres. In fact, the photographs of orbits may resemble bubble-chamber pictures. In a comparison of the two instruments the bubble chamber has the disadvantage that it cannot be triggered. Both devices may register many simultaneous particles. In the streamer chamber a simple mixture of noble gases (neon and helium) is normally used—at room temperature and atmospheric pressure.

The streamer chamber has been called by many names. For example, the original papers referred to the "track spark chamber" (Chikovani *et al.*, 1964), the "isotropic" spark chamber" (Gygi and Schneider, 1966) and the "streamer chamber" (Dolgoshein *et al.*, 1964). We shall use the expression "streamer chamber" as a generic name for all apparatus in which the discharge is severely limited in duration, in spite of the fact that the description is not always strictly accurate.

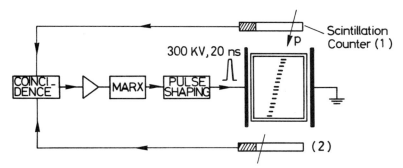

Fig. 6.1.1.1 Elements in a streamer chamber system.

The basic instrumentation for a streamer chamber is shown in Fig. 6.1.1.1. It differs from the wide-gap spark chamber system mainly in the addition of a pulse-shaping network to reduce and control the duration of the applied field. Typically, a field of about 20 kV/cm has to be applied to the sensitive gas volume for about 20 ns. Methods for achieving this are discussed in section 6.2.

Three distinct regimes have been employed in streamer-chamber experiments. They are shown in Fig. 6.1.1.2 where it

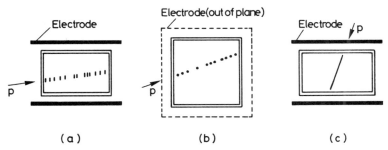

Fig. 6.1.1.2 The three distinct regimes for streamer operation. The appearance of a track is indicated when viewed from the position of the reader. (a) Isolated streamers – side view. (b) Isolated streamers – end view. (c) Merged streamers – side view.

is seen that the differences concern the mutual orientation of the tracks, the electric field and the photographic axis. Our task in the next few sections will be to outline the characteristics of each regime.

6.1.2 Isolated streamers—side view

The basic configuration is a box chamber enclosing the sensitive gas, sandwiched between two large electrodes. The conventional spark-chamber gas, neon plus helium, is customarily used. On triggering, a rectangular pulse is applied to one plate, the other being earthed. During the pulse, some of the electrons liberated by the traversing particle initiate avalanches and grow into streamers. At the end of the pulse, the field disappears, and the streamers remain suspended in the chamber as small stationary clouds of electrons and ions. A camera situated orthogonally to the field and whose shutter must have been opened before the event, will record the light emitted by the recombining ions. Each streamer will appear as a cylinder with, for example, a length of 7 mm and a diameter of 1 mm; and the track will be indicated approximately by a line through the centres of the cylinders.

The photographs in Fig. 6.1.2.1 show side-views of streamer trails arrested at different stages in their growth. They also indicate the difficulty of locating the precise trajectory. On the one hand, it is not easy to pin-point the centre of the streamers for the luminous region is diffuse and varies from streamer to streamer. And, in any case, the centre need not indicate the exact site of ionization. The relative position of this site will depend partly on the nature of the initial avalanche which itself will depend on the pulse applied to the chamber. The finite rise-time of this pulse may cause an appreciable drift prior to gas multiplication and also affect the size of the avalanche. It is clear that this operating regime is far from ideal.

6.1.3 Isolated streamers—end view

The technique in this regime is to employ transparent electrodes and to aim the photography along the line of the electric field. The cylindrical streamers will then be viewed end on (see Fig.

(a)

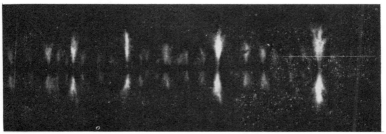

(b)

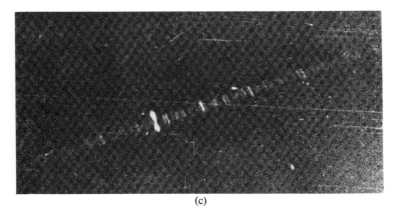

(c)

Fig. 6.1.2.1 Photographs showing streamers at different stages in their growth, viewed from the side: (a) with the field applied for about 100 ns, the streamers have grown to reach the electrodes (Chikovani et al., 1964); (b) the streamer growth has been arrested by applying the field for about 60 ns (Chikovani et al., 1964); (c) in this case, the field was applied for only 35 ns with the result that the streamers are only 7 mm in length. The trajectory of the ionizing particle becomes quite clear. (Dolgoshein et al., 1964)

6.1.3.1). The centres of the well-defined projections of the streamers will indicate the true track. The finite radii of the streamers are merely due to lateral diffusion and are not influenced by the field. Assuming that the streamers emit their light isotropically, the reduced luminous area in this projection makes photography easier—the dots will be brighter.

Fig. 6.1.3.1 An end (projected) view of isolated streamers along a 20 cm cosmic ray track obtained with a nominal field of 20 kV/cm. (Rice-Evans and Mishra, 1969)

This mode of photography is to be preferred to the side view. On the other hand, it may be useful to take both views simultaneously, in which case a three-dimensional estimate of the track will be obtained. The adroit use of mirrors will permit the recording of both views on one negative. The most sensible approach is to use the end-view mode with a stereo angle of about 20°, i.e. not 90°.

6.1.4 Merged streamer regime

The third regime is one in which the particle trajectory lies within about 30° of the electric field. In this case the individual

GENERAL PICTURE

avalanches, or streamers, merge to form a continuous plasma along the line of the trajectory. Stereo photography is easy from the sides of the chamber. This plasma is the same as that initially created in the wide-gap spark chamber (see section 5.2). No current is drawn directly from the electrodes however; rather the gas multiplication alone is enough to provide a luminous trail. This approach has the advantage that spurious discharges emanating from the electrodes are eliminated. Figure 6.1.4.1 shows some merged streamer trails along cosmic ray trajectories; in a 2 × 20 cm gap chamber.

Fig. 6.1.4.1 The merged streamer regime. (Rice-Evans and Mishra, 1969)

6.1.5 General picture

Following the passage of a particle through a streamer chamber and the application of the high-voltage pulse, the primary ionization electrons released by the traversing particle initiate avalanches. These transform into streamers when their internal space-charge field becomes comparable with the applied field. A few nanoseconds later they become visible and the field may be cut off. A camera will record a row of luminous streamers localized along the particle trajectory.

The actual number of streamers will be less than the number of primary electrons. This is because the statistical fluctuation in the growth of the avalanches means that some mature before others. In this event, the space-charge field of the first-developed streamers will inhibit the growth of avalanches in the regions immediately adjacent to them and thus prevent their development into streamers. In addition, two or more primary electrons may participate in essentially the same avalanche. Such joint avalanches are likely to develop in advance of the others, which is good because they will probably lie closer to the true particle trajectory.

6.1.6 Aspects of timing

Attention must be paid to the delay between the passage of the particle and the application of the high-voltage pulse. The delay may be adjusted electronically from a minimum of about 200 ns up to several hundred microseconds. In reasonably pure neon the primary electrons will remain unattached for long periods; Chikovani et al. (1964) mention that in neon with 0·1% air the average lifetime of an electron before capture is about 1 ms, and that the decrease in 200 μs is merely 20%. This effect is therefore negligible with a typical delay time of less than 1 μs.

In experiments where a streamer chamber has to be rapidly recycled, the longevity of unattached electrons could prove an embarrassment. It is impractical to provide a clearing field to sweep away the electrons on account of the large gap (cf. multiplate chambers). The addition of small concentrations of electronegative gases offers a better solution. Bulos et al. (1967) report that freon-12, ethylene and methane were all found to be effective in reducing the memory time. But they also mention that undesirable side-effects were found with SO_2. For example, with fourteen parts per million of freon-12, the number of observed streamers/cm dropped from 3 to 0·5 as the delay was increased from 150 ns to 7 μs. They concluded that memory times of 1 μs could easily be achieved.

During the delay (t_d) prior to the application of the high-voltage pulse, the ionization electrons diffuse. This diffusion will be away from the true trajectory of the particle. One would therefore expect that a track would be indicated by a trail of streamers showing a scatter with an r.m.s. deviation given by $\sigma_{\text{diff}} = \sqrt{(2Dt_d)}$ where D is the electron diffusion coefficient. The scatter distribution has been observed and measured by Chikovani et al. (1964). They found values for σ of 0·3 mm and 1·1 mm for delays of 1 μs and 11 μs respectively. These were only half the values calculated with $D = 2 \cdot 10^3$ cm²/sec representing the diffusion of single electrons in neon. The authors concluded that the streamers are in fact initiated by clusters of two or more electrons.

Another effect of increasing the delay is that the observed number of streamers actually increases. This is a consequence of the distinction between primary and total ionization (see section 1.2.2). At small delays the secondary electrons are so close to the primary electrons that they participate in the same avalanches. With longer delays, however, the secondary electrons have a chance to diffuse to a distance where they may initiate their own streamers. This effect is of importance in the measurement of ionization loss and will be discussed in section 6.6.

6.2 HIGH-VOLTAGE PULSE SHAPING

It has been the ability of scientists critically to control the shape of the high-voltage pulse, that has led to the successful realization of streamer chambers. Unless extraordinary precautions are taken, the pulse from the Marx generator has, in practice, a relatively long rise-time. As a result, a few networks have been designed to improve the pulse, and these will be described in the next sections. The broad requirements are a pulse of say 300 kV, rising in a few nanoseconds, of duration about 20 nsec and with a rapid fall off. At these high frequencies attention must be paid to the application of the pulse, so as to ensure that all parts of the sensitive gas volume experience the same field conditions. It should be noted that in a streamer chamber even though the discharge is isolated from the electrodes by an insulator, the discharge will act as a shunt across the electrodes. In the case of Falomkin et al. (1967) this effect diminished the amplitude of the pulse by a factor of 2 in 25 nsec.

6.2.1 Series and shunt spark gaps

The first technique to shorten the rise-time of the Marx output before application to the streamer chamber, was introduced by Chikovani, Roinishvili et al. (1965). They placed a series spark gap (S_1 in Fig. 6.2.1.1) in between the Marx and the chamber, and also a low-inductance shunt capacity (c) at the output of the Marx (to augment the stray capacitance). The result was that when the Marx fired, the pulse first charged this capacity (relatively slowly). When the maximum voltage was approached, the spark gap S_1 broke down, and only then was the voltage

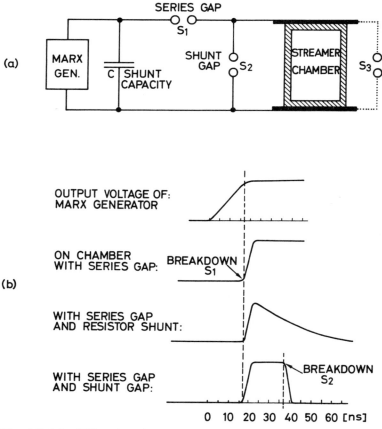

Fig. 6.2.1.1 HV pulse shaping with series and shunt spark gaps.

SERIES AND SHUNT SPARK GAPS 205

transferred to the chamber. The figure shows this sharpening of the applied pulse: the rise-time can be as short as 1 nsec and is limited mainly by the inductance of the spark gap.

The duration of the voltage on the chamber electrode may be controlled merely by connecting it to earth through a resistor of chosen value. The charge will then leak away exponentially at a rate defined by the effective time constant. A better way, however, is to use a shunt spark gap S_2 across the chamber. The separation of the gap may be chosen so that the gap will break down after a prescribed time. With this method, one expects the voltage pulse on the chamber to have a flat top and then drop sharply to zero.

In practice under some conditions the generator may continue to feed what is basically an oscillatory circuit, and the voltage may rise considerably above the Marx rating (see section 5.3.1).

Energy requirements ($\frac{1}{2}CV^2$) suggest that the effective Marx capacity should be greater than the shunting capacity, which itself should be greater than the chamber capacity. Thus in the early exquisite work of Chikovani *et al.* (1965), the respective capacitances were 1000 pF, 160 pF and 22 pF. They employed nitrogen at 5 atmospheres in their series spark gap, but their shunt gap was sometimes in air, sometimes in pressurized nitrogen. The statistical variation in the time of shunt-gap breakdown can be troublesome. Chikovani found it desirable to "prepare" the gap by feeding a pulse taken from an early Marx stage, to a small third electrode in the shunt gap. This caused a tiny spark to the earth electrode 60 nsec prior to the main discharge.

When short pulses are applied to a chamber, there is always the problem of reflections from the far end. Chikovani *et al.* found that they achieved the best results by shunting the far end with additional spark gaps—rather than by shunting with resistance.

Gygi and Schneider (1971) have achieved subnanosecond rise- and fall-times with the arrangement shown in Fig. 6.2.1.2. A pulse-forming network (PFN) is charged by a Marx slowly through an inductance L up to a high voltage. The PFN constitutes a strip line which may be discharged into the matched load with a series switch. When the switch fires, 200 kV on the PFN will result in a square 100 kV pulse on the load for twice

206 STREAMER CHAMBERS

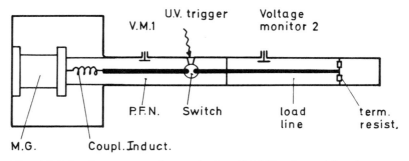

Fig. 6.2.1.2 A generator for producing 100 kV pulses with subnanosecond rise and fall times. The switch consists of four spark gaps spread across the electrode. (Gygi and Schneider, 1971)

the transit time (τ) in the PFN. This assumes that the coupling inductance appears as an open circuit during the pulse-forming —a condition that is fulfilled if $L \gg Z_0\tau$, i.e. 2 μH here.

To ensure the remarkable timing characteristics across a 25 cm wide output electrode, the PFN consisted of four separated parallel bars electrically connected at the inductance end. The pulse is initiated across the load by firing four parallel spark gaps spread across the 25 cm electrode. The spark gaps are triggered with a precision of 0·1 ns by an ultra-violet pulse derived from an auxiliary spark gap (5 atm of N_2 or A, with a corona needle) driven by another 200 kV Marx. With this arrangement square 4 ns, 100 kV pulses were produced on a 20 ohm load with 0·6 ns rise- and fall-times. This example approaches the limits imposed by spark-gap breakdown, e.g. the probability that an electron will be liberated by a field of 10^7 V/cm from a 0·1 cm² metallic surface with a work function of 4 eV, within 40 ns, is 10^{-2}.

6.2.2 Blumlein line

When large (say, 2 m) streamer chambers are being designed, it must be remembered that the chamber dimension is of the same order as the pulse length. It is therefore vital to consider the chamber as a transmission line, and to ensure that the correct conditions for pulse termination exist at the end of the chamber away from the generator. Bulos *et al.* (1967) have explained that when a chamber is designed as a transmission line, then it can be incorporated into a Blumlein line, i.e. a network that may subject the chamber to a well-defined rectangular high-voltage pulse.

To understand the action of a Blumlein line one needs to know that a voltage pulse travelling down a transmission line will at the far end be

(a) reflected unattenuated with the opposite polarity if the end be short-circuited ($Z = 0$).

or (b) reflected unattenuated with the same polarity if the end be open-circuited ($Z = \infty$).

or (c) totally absorbed (with no reflexion) if the end be terminated with a resistance with the value of the characteristic impedance of the line ($Z = Z_0$).

The application of the Blumlein voltage doubling circuit to radar has been described by Ratsey (1946) and Wilkinson (1946). In Fig. 6.2.2.1(a) it is seen that two lines are effectively charged in parallel. When the spark gap at one end fires, a current wave of magnitude $I = V/Z_0$ flows in the line, discharging all the condensers to zero voltage, and building current in the inductances. In the radar case, the load is a magnetron which is open circuit at the instant the pulse arrives, so the wave is reflected and charges all the left-hand condensers to $-V$. The voltages then add, to give effective voltage $2V$, which is divided between the series networks and the matched load, to give a voltage $-V$ across the load for twice the transit time of the networks.

In the streamer-chamber application, the action of the Blumlein line may be understood by considering the three-electrode arrangement in Fig. 6.2.2.1(d) (Crouch and Risk, 1971). Suppose that the central plate is charged, relatively slowly, by the Marx generator to the high-voltage V, at which value the spark gap on the left breaks down ($t = 0$). When this happens a pulse of amplitude V, and polarity opposite to the original voltage, propagates down the line to the right. When the pulse reaches the end of the centre plate (at the time τ) it sees a total impedance of $3Z_0$—i.e. the series sum of Z_0 for the upper line plus $2Z_0$ for the output line to the right. At this junction part of the wave will be reflected back along the lower left-hand line. This part

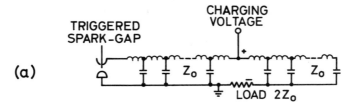

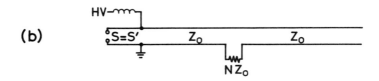

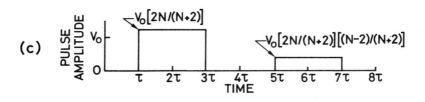

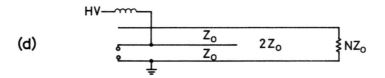

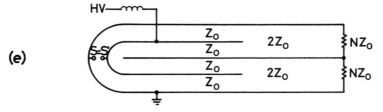

Fig. 6.2.2.1 The Blumlein line and its operation (see text).

will have amplitude V_R and polarity the same as the incident wave. Thus the voltage across the lower left-hand line at the junction will be $(V_{\text{orig}} - V_{\text{incid}} - V_R) = -V_R$, where if I be the incident current

$$V_R = Z_0 I_R = Z_0 \left(\frac{3Z_0 - Z_0}{3Z_0 + Z_0}\right) I = \tfrac{1}{2} I Z_0 = \frac{V}{2}$$

Part of the wave will be transmitted:

$$V_T = (3Z_0) I_T = 3Z_0(I - I_R) = 3Z_0 I/2 = 3V/2$$

Of this transmitted voltage, $V/2$ is across the upper line (Z_0) and V is across $2Z_0$, i.e. the output electrodes (where there was zero original voltage). We see that the output voltage equals the initial-charging voltage. Thus a single-gap streamer chamber incorporated between the output electrodes will experience a pulse of amplitude V.

After the moment τ, the wavefronts in the upper and lower lines move to the left. The lower wave is reflected at the spark gap ($Z = 0$) with inverted polarity, and the upper wave is reflected at the open end ($Z = \infty$) with the same polarity. The result is that at the moment 3τ, they both arrive back at the junction with opposed voltages and thus drive the output line to earth. All succeeding reflexions proceed similarly to cancel. The pulse that the streamer chamber experiences therefore lasts for a time 2τ ($3\tau - \tau$). If the far end of the streamer-chamber electrodes be terminated with the characteristic impedance of the chamber ($2Z_0$) no reflexions should occur.

The general case with a load NZ_0 has been examined by Bulos et al. (1967). In Fig. 6.2.2.1(b), by the same reasoning, the transmitted voltage at the junction is given by

$$V_T = Z_0 I \left(1 - \frac{(N+1)Z_0 - Z_0}{(N+1)Z_0 + Z_0}\right) = V \left(\frac{2N}{N+2} + \frac{2}{N+2}\right)$$

where V is the charging voltage. This voltage is shared: $V2N/(N+2)$ is the voltage across the load, $2V/(N+2)$ is the voltage proceeding towards the open end of the line, and $VN/(N+2)$ is the amplitude of the reflected pulse. After reflections, the pulses arrive back at the junction at time 3τ where they cancel the voltage across the load. In the unmatched case however ($N \neq 2$), further reflexions do not cancel, with the result that at time 5τ a further pulse of amplitude

$$V\left(\frac{2N}{N+2}\right)\left(\frac{N-2}{N+2}\right)$$

appears across the load and persists until 7τ (Fig. 6.2.2.1(c)).

The matched value ($N = 2$) is chosen for the streamer chamber, in which case one expects a single rectangular pulse across the electrodes. Figure 6.2.2.1(e) shows a five-element Blumlein —which is equivalent to a pair of three-element Blumleins like Fig. 6.2.2.1(d). It is this five-element configuration that Bulos et al. have used to drive their double-gap chamber, with the central element as the central electrode.

It is clear that if the chamber electrodes are to be part of the Blumlein transmission line, care has to be taken over the question of impedance. For a large chamber with $2l \geqslant c(2\tau)$ [l is the chamber length and 2τ the pulse length], the characteristic impedance is

$$Z = \frac{377}{\sqrt{(\varepsilon_{\text{rel}})}} \frac{d}{w+d}$$

where d and w are the gap spacing and width respectively. This is the standard for a strip line, where the d in the denominator is included to account for edge effects. Thus, for example, the SLAC chamber with electrodes (250×150 cm^2) constituting two 30 cm gaps in parallel, has a combined impedance of about 30 ohms. The Blumlein line dimensions and the terminating resistors must be chosen to match the chamber. At SLAC, the open elements of the Blumlein line were immersed in oil and the terminating resistors were ten carbon rods spaced across the width of each gap. Figure 6.2.2.2 shows a diagram of the system.

Analogue methods with lowly conducting paper were used to match the Blumlein line and chamber impedances. If a long chamber is not terminated, the original pulse is reflected back with the same polarity. This can mean that the input side of the chamber sees the pulse for a longer duration, and the far end experiences twice the field! Termination of the SLAC chamber with an impedance 35% lower than the characteristic im-

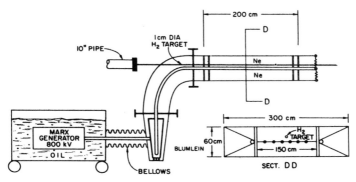

Fig. 6.2.2.2 Diagram of the SLAC Blumlein. (Odian, 1966)

pedance had a marked effect on the development of streamers along the track.

An awkward point about the Blumlein operation is that prior to sparking, while it is charging, it is liable to transmit a "prepulse" to the chamber. This prepulse is opposite in polarity to the proper output pulse and it may reach a comparable amplitude if the rise-time of the Marx is fast. One cure is to connect the Marx to the Blumlein through an inductance which increases the charging time (to, say, 100 ns). Of course, this can be most deleterious if one needs to minimize the delay in pulsing the chamber. Bulos et al. have employed a bridge circuit to remove the prepulse.

A compact Blumlein line capable of producing 5 ns long, 240 kV pulses has been built by Crouch and Risk (1971) at Maryland (Fig. 6.2.2.3). By using glycerine ($\varepsilon = 44$) as the dielectric medium they were able to reduce the impedance to 14·5 ohm. Another possibility was to use deionized water ($\varepsilon = 80$). An important consideration is the high-frequency characteristics of the dielectric; Table 6.2.2 suggests that for short lines ($\leqslant 30$ cm) rise-times down to a few nanoseconds are possible. Typically, the spark gap contained about 100 p.s.i. of SF_6. With a twelve-stage Marx generator, of effective capacity 975 pF feeding the 475 pF Blumlein line (+ gap); with 30 kV per stage, the Blumlein line output was

$$(975/1450)(12 \times 30) = 240 \text{ kV.}$$

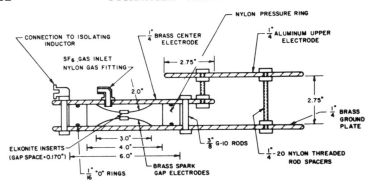

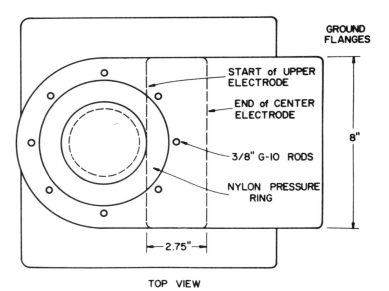

Fig. 6.2.2.3 Constructional details for a compact glycerine Blumlein. (Crouch and Risk, 1971)

6.2.3 Measurement of high-voltage pulses

The measurement of short high-voltage pulses presents a number of problems. An ultrafast oscilloscope (e.g. Tektronix 519) is essential and the pulses must be faithfully attenuated

TABLE 6.2.2

Dielectric	Temp.	Frequency (CPS)	1×10^7	1×10^8	3×10^8	3×10^9	1×10^{10}
Ethelene glycol	25°C	$\varepsilon'/\varepsilon_0$ $\tan \delta \times 10^4$	41 80	41 450	39 1600	12 10,000	7 7800
Water, conductivity	25°C	$\varepsilon'/\varepsilon_0$ $\tan \delta \times 10^4$	78·2 46	78 50	77·5 160	76·7 1570	55 5400

High-frequency characteristics of glycerine and conductivity water.
$\varepsilon'/\varepsilon_0$ is the relative dielectric constant and $\tan \delta$ is the dielectric-loss tangent. The conductivity and attenuation length α^{-1} are given by $\sigma = 0.56(\varepsilon'/\varepsilon_0) f \tan \delta \times 10^{-12}$ mho/cm (f is frequency in cycles per sec.); and $\alpha^{-1} = \lambda_0 (2\pi^2 \varepsilon'/\varepsilon_0)^{\frac{1}{2}} (\sqrt{(1 + \tan^2 \delta)} - 1)^{\frac{1}{2}}$ meters (λ_0 is free space wavelength). (Crouch and Risk, 1971; von Hippel, 1954.)

before measurement. In addition, precautions have to be taken to shield the whole streamer-chamber assembly against r.f. radiation; and mains filters are necessary to prevent conducted feedback. Short low-inductance ground loops are advisable wherever practical.

Both resistive and capacitive attenuating probes have been constructed. Figure 6.2.3.1(a) shows a resistive probe with an attenuation factor of 53. The input resistance of 1000 ohms is

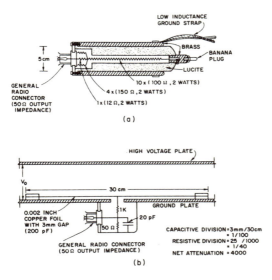

Fig. 6.2.3.1 HV measuring probes used at SLAC: (a) resistive probe; (b) capacitive probe (Bulos et al., 1967)

distributed along a string of ten internal carbon resistors, and the impedance looking back into the output is 50 ohms, to match the 50 ohm coaxial cable. The probe is capable of measuring pulses of up to 25 kV with a rise-time of 1 nsec. Higher voltages may be studied by connecting the probe across a carbon resistance near the earthed end of a resistance chain connected to the high-voltage electrode.

Capacitative probes allow a better measurement, but they are more difficult to realize. In the case when the pulse is fed into the chamber through a transmission line, a capacitative divider may be obtained by placing a flat probe electrode inside the line and just above the earthed plate. The SLAC arrangement is shown in Fig. 6.2.3.1(b). A thin copper foil is stretched between two perspex rings, and an output lead taken through a hole in the earth electrode. The capacitative division is 3 mm/30 cm = 1/100, the additional resistive division 25 ohms/1000 ohms = 1/40, making a total attenuation of 4000. Bulos et al. noted that extra shielding of the signal cable was necessary to obtain clean pulses. Clearly visible traces on the 519 showed that the 600 kV pulses applied to the chamber had a 4 nsec rise and a duration of 12 nsec.

Ringing may occur in capacitative dividers if the chosen dielectric of the probe capacitance is different from that of the transmission line. This is because of the different wave velocities in the two dielectrics. In the SLAC case, the difficulty does not arise because the dielectric is air in both parts. On the other hand, Gygi and Schneider (1964) decided to achieve an attenuation of 10^4 (200 kV to 20 V) entirely with a capacitative probe. To do this in their geometry (11 cm gap) the probe electrode had to be separated from the bottom plate by less than 0·1 mm. Furthermore, to prevent the probe appearing as an obstacle it had to be extremely thin. To achieve this a 4.10^{-6} cm thickness of gold was used on a 4 cm long, 9.10^{-2} cm thick mica ($\varepsilon = 6·5$) support. This metallic thickness had to be smaller than the skin depth of the highest frequency ($\omega = 10^{10}$ Hz), i.e. less than $\sqrt{(2\rho/\omega\mu)} = 6.10^{-5}$ cm, where ρ is the resistivity of the film and μ the permeability of the support. This ensured that the travelling wave was negligibly perturbed. The chosen dimensions also satisfied requirements concerning the resistivity of the film that came from an analysis of the equivalent circuit (Fig. 6.2.3.2).

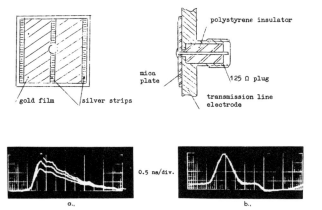

Fig. 6.2.3.2 Construction of divider electrode by Gygi and Schneider (1964). Left trace shows the Marx pulse; right, the pulse after clipping.

6.3 EXAMPLES OF STREAMER CHAMBERS

6.3.1 Bedford College streamer chamber

This is an example of a low budget chamber that might be built in a modestly sized university department (Rice-Evans and Mishra, 1969). Physically, the chamber is composed of 61 × 61 cm) electrodes sandwiching perspex boxes that enclose a 20 × 20 × 20 (cm) volume of sensitive gas (Fig. 6.3.1.1). The 12 mm perspex walls have "O" ring grooves milled in them to allow both a flexible arrangement of electrodes and evacuation prior to filling with gas. The electrodes, whose dimensions were chosen to ensure a reasonable uniformity of electric field within the chamber, were constructed from both aluminium (6·4 mm) plate and also wire grids (0·1 mm diameter stainless-steel wire, 2 mm spacing, with ends soldered to copper straps). The latter were wound under tension on 61 × 61 × 0·64 (cm) perspex formers with 20 × 20 (cm) holes in the centre to correspond with the sensitive region. To allow "O" ring sealing, perspex spacers were cemented so that the wires were sandwiched in the cement. In some configurations the rear interior wall was painted matt black to preclude reflexions. An important feature

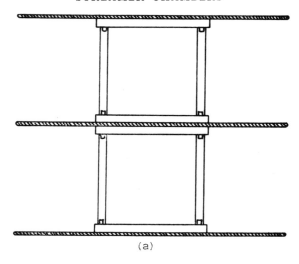

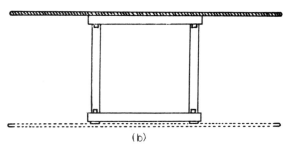

Fig. 6.3.1.1 Sketch of Bedford chamber construction: (*a*) for the merged streamer regime with aluminium electrodes; (*b*) for photographing isolated streamers in the projection mode through a transparent wire-grid electrode. (Rice-Evans and Mishra, 1969)

is that 12 mm of perspex separates an electrode from the sensitive gas. Figure 6.3.1.2 shows a photograph of the chamber.

The trigger amplifier is shown in Fig. 3.1.1.2. Two volt coincidence pulses from the particle counters are amplified in a valve circuit in which the CV 4082 produces a sharp 6 kV output pulse. This pulse triggers the first spark gap, in which barium titanate is used to facilitate sparking (Lavoie *et al.*, 1964). Breakdown of the 22 kV at this gap transfers the charge from

Fig. 6.3.1.2 The Bedford College streamer chamber

capacitance 2500 pF to the first gap of the Marx generator (Fig. 6.3.1.3).

This generator consists of four cylindrical rapid-discharge capacitances mounted coaxially with the stainless-steel spark electrodes. Each capacitance charges up to 60 kV. When triggered, the spark gaps breakdown as they become over voltaged and the final output pulse has a nominal magnitude of 240 kV. To encourage rapid breakdown, small (2 pF) polystyrene dielectric by-pass condensers are connected as shown (Gygi and

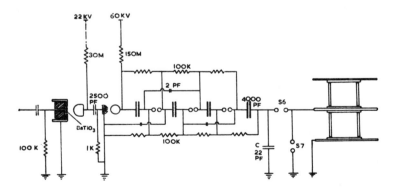

Fig. 6.3.1.3 Four-stage Marx generator. A negative 6 kV pulse triggers the first spark gap, the Marx fires, and a nominal, shaped 240 kV pulse is applied to the chamber. (Rice-Evans and Mishra, 1969)

Schneider, 1964); and the components and interior walls of the generator are painted white to reflect the ultra-violet light from one gap to the others.

A series-shunt spark-gap arrangement shapes the pulse. An auxiliary stray capacitance C (44 pF), constructed from aluminium foil and polystyrene foam, is charged by the Marx output and on breakdown of the series gap, a rapidly rising pulse appears on the chamber electrodes. The spark gaps consist of 2 cm diameter stainless-steel ball electrodes in air, and they are adjusted to give optimally precise and visible tracks. The pulse lengths are typically 30–50 ns, and the overall delay between the passage of the particle and the application of the high voltage is about 400 ns.

In the electronics, valves have been used as they have been found to be more reliable in the proximity of high-voltage surges. Although the Marx has worked well, corona problems do appear when one uses 60 kV d.c. in air. They can usually be overcome by smoothing surfaces and applying silastomer rubber solution, but experience would suggest that more stages and a lower voltage might be a better recipe.

Photographs of cosmic-ray streamer tracks are shown in Figs. 6.1.3.1 and 6.1.4.1. In the case of the merged streamer regime, fine continuous tracks are obtained up to about 30° with respect to the field. In the end view (projection mode) tracks are clearly delineated at 10 kV/cm, but an improvement is obtained if a stronger field (20 kV/cm) is employed—by using a narrower chamber. The picture of the cosmic shower (Fig. 6.3.1.4) indicates the chamber's capacity for recording many simultaneous tracks.

In a study on the multiple scattering of muons in lead (Rice-Evans, Hassairi and Mishra, 1973) the merged streamer regime was adopted. It turned out that the breakdown time of the shunting spark gap fluctuated sufficiently to cause a variation in the brightness of the tracks, from one trigger to the next. As a result, an alternative system for attenuating the applied field was employed (Rice-Evans and Hassairi, 1973). An empirical adaptation of Lecher's wires was used; the electrodes were connected to a parallel wire transmission line of about 3 metres length, that was shorted at the far end. The resulting tracks are more diffuse than in the case of the spark gap—but experience

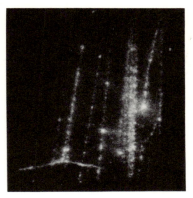

Fig. 6.3.1.4 A cosmic ray shower in the Bedford streamer chamber. (Rice-Evans and Mishra, 1969)

has shown them to be constant in intensity (Figs. 6.3.1.5 and 9.4.3.1).

6.3.2 The SLAC streamer chamber

A splendid large streamer chamber has been built at the Stanford Linear Accelerator Centre. It was designed (by Bulos, Odian, Villa and Yount, 1967) to sit inside a large magnet (Fig. 6.3.2.1). The double-gap chamber consists of two cells, with polyurethane foam walls, each of 30 cm depth, 150 cm width and 230 cm length. The cells have mylar windows which are separated from the electrodes without by $\frac{1}{8}$ in. perspex rods.

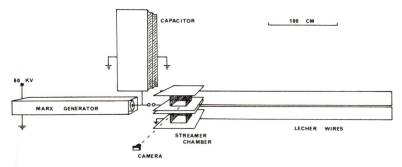

Fig. 6.3.1.5 High-voltage pulse-shaping elements for a streamer chamber. (Rice-Evans and Hassairi, 1973)

The electrodes are of 0·25 mm wires, spaced 1 cm apart and supported on a ¼ in. thick window-frame. The whole electrode structure has a length of 4 m, a width of 3 m and a height of 70 cm. (N.B. The skin depth (cm) is equal to $5 \times 10^3 \, (\rho/f)^{\frac{1}{2}}$, where ρ is the resistivity (ohm/cm), f is $1/(4t)$ and t is the risetime of the pulse; and it is important to check that electrodes are sufficiently thick to prevent them assuming a large high-frequency resistance.)

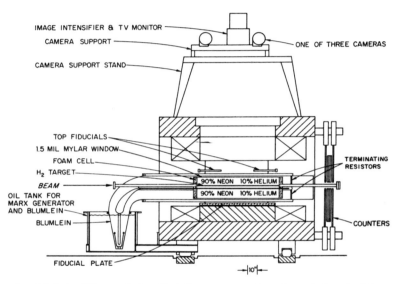

Fig. 6.3.2.1 View of the SLAC streamer chamber sitting inside a magnet. For a photoproduction experiment, a collimated bremsstrahlung beam passes down a half-inch diameter hydrogen gas target in the upper cell. (Bulos *et al.*, 1967)

The chamber is continuously flushed with a neon helium mixture at a rate of 500 litres/hour. The gas is recycled through a purifier which keeps the impurities down to 0·2% which is well below the 0·5% limit for bright streamers.

A special feature for photoproduction experiments is the placing of a ½ in. diameter cylindrical hydrogen gas target down the length of one of the cells and just above the central electrode. Any charged particles created when a collimated bremsstrahlung beam passes down this tube are presented with essentially a 4π

sensitive geometry. The electrodes are transparent and so tracks in both cells can be photographed from above.

The pulsing system is capable of producing a 600 kV pulse across the electrodes for a 10 nanosecond duration. The overall delay after triggering has to be less than 1 μsec, and the apparatus is normally pulsed at the 10 Hz accelerator rate. The energy required across 23 ohms is 100 joules. To achieve this, the researchers chose a thirty-four-stage Marx generator coupled to a Blumlein line (see section 6.2.2).

The photography is performed with three stereoscopic cameras, all pointed in the direction of the field. The streamers are therefore viewed "end on" and particle trajectories are clearly indicated. In one K_{OL} decay experiment 1·5 million photographs were taken (Villa, 1970)—which is a measure of the reliability of the design (Fig. 6.3.2.2).

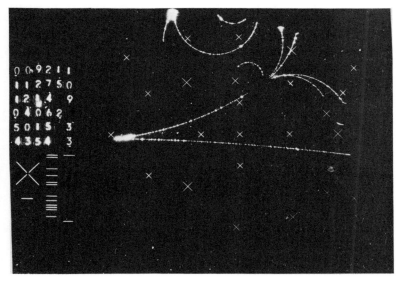

Fig. 6.3.2.2 An event in the SLAC streamer chamber. (Odian, 1970)

6.3.3 The CERN avalanche chamber

The technical virtuosity of Gygi and Schneider led to the development in 1966 of a most elegant little chamber. Employing their Marx generator (see section 5.3.1) they were able to apply approximately triangular 300 kV pulses, of only 5 ns base width, to a chamber of 10 × 10 × 11 (cm) volume. This short pulse duration arrested the avalanches before they grew into streamers. Relatively little light ($\sim 10^8$ useful photons) was emitted by the 1 mm long avalanches, but it was sufficient satisfactorily to be recorded with an image intensifier.

The compact dimensions of the Marx generator enabled it to be placed within an 11 cm transmission line. The chamber electrodes, one aluminium, the other conducting glass, were built into this transmission line. The side walls were of perspex. The isolated avalanches along the track could then be viewed both from the side through the perspex, and parallel to the field through the conducting glass. The 2 ns rise-time of the Marx pulse meant that a series spark gap was unnecessary. The long pulse tail was clipped with a pressurized nitrogen shunting spark gap in which preionization was produced by means of a corona discharge.

An important feature of the avalanche chamber is that all the avalanches resulting from the primary ions should be observed. This contrasts with a chamber operated in the streamer mode. In the latter case, the growth of smaller avalanches lying immediately alongside the first-developing streamers is inhibited by the space-charge field. The number of streamers may therefore be less than the number of primary avalanches. Figure 6.3.3.1 shows photographs of tracks taken with an image intensifier. The short avalanche length in the field direction is to be noted. A minimum-ionizing particle produces about 9·5 avalanches per cm in 70% neon/30% helium. This regime is, therefore, capable of measuring primary ionization, and this will be discussed in section 6.6. A typical photograph taken in a CERN experiment is shown in Fig. 6.3.3.2. See also section 6.6.3.

6.3.4 The DESY (Hamburg) streamer chamber

A double-gap streamer chamber has been built to sit inside an 18 kGauss bubble-chamber magnet at DESY. The earthed electrode completely surrounds a perspex box (100 × 60 × 48 cm^3) and hence provides good r.f. shielding (Fig. 6.3.4.1). The high-voltage wire-mesh plane (75% transparency) lies

THE CERN AVALANCHE CHAMBER

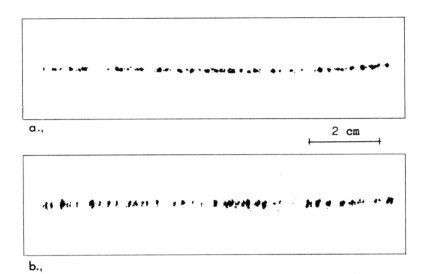

Fig. 6.3.3.1 Avalanches registering a particle track, photographed with an image intensifier: (*a*) viewed parallel to the field; (*b*) viewed perpendicular to the field. (Gygi and Schneider, 1966)

Fig. 6.3.3.2 A typical photograph taken with the avalanche chamber in a CERN experiment. (Schneider, 1969)

centrally in the box. The earthed electrode on one side contains a wire-mesh window (also 75%) to allow photography of the streamers in the projection mode—in both gaps (Eckardt and Ladage).

Inside one of the gaps of the chamber a liquid hydrogen target is incorporated for photoproduction experiments. This cylindrical target (15 mm diameter, 90 mm long) is isolated from the chamber gas by a vacuum jacket enclosed in a perspex foam tube of 10 mm wall thickness (0·06 g/cm^3). This tube stretches between the electrodes and prevents breakdown along the target walls.

The high-voltage pulse derived from a ten-stage Marx generator, with a capacity of 8 nF per stage and operated at ± 35 kV. A coaxial Blumlein line geometrically matches the Marx output to the cross-section of the chamber (impedance 27 ohm), and shapes the pulse to 10 ns FWHM, with 3 ns rise- and fall-times.

Stereophotographs at 18° were taken with three cameras using 10,000 a.s.a. film. The dead time (~ 0.5 sec) of the system was governed by the camera transport and the recharge time of the Marx. Typically, a repetition rate of 1 Hz is used, with a memory time of 2 μs (see section 9.1.4).

6.4 RECORDING IN STREAMER CHAMBERS

6.4.1 Brightness of streamers

The luminosity of a streamer is essentially a matter of recombination and the emission will therefore be spectral rather than continuous. A calculation of the variation of the light output along the axis of a single avalanche (see Fig. 1.4.3.4) shows that most of the light is emitted from the head of the avalanche. Even after transition into a streamer, the intrinsic brightness remains low: Davidenko *et al.* (1969) estimate an emission of 10^{10}–10^{12} photons/mm^2.

Bulos *et al.* (1967) have made a rough estimate of how the brightness varies with the applied field. Their treatment is restricted to the case of an avalanche that has attained the Raether–Meek criterion of 10^8 electrons, and does not entertain

THE DESY (HAMBURG) STREAMER CHAMBER

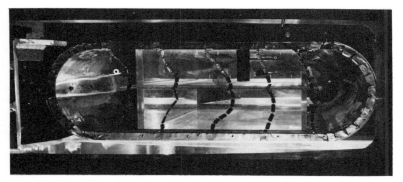

Fig. 6.3.4.1 View into the DESY streamer chamber. (Ladage, 1969)

Fig. 6.3.4.2 View into the DESY streamer chamber from the camera side. (Ladage, 1969)

photoionization. First, the following equations produced in Chapter 1 must be recalled:

$$dn = \alpha n\, dx$$
$$n = \exp(\alpha x)$$
$$x = ut = \mu E t$$
$$r^2 = 4Dt$$
$$D \propto E^{\frac{1}{2}}$$
$$x_c = 20/\alpha$$

Assuming that the number of photons created is proportional to the electron concentration, one may write

$$dp = \delta n\, dx$$

where δ is the number of photons produced by one electron per cm.
Then

$$dp = (\delta/\alpha)\, dn = F\, dn.$$

This ratio F, of rates of photon and electron production, is found experimentally to vary as $1/E$. Taking, roughly, α proportional to E,

$$x_c \propto 20/E \propto \mu E t_c$$

and one concludes

$$E^2 t_c = \text{constant}$$

(but note that the length of the avalanche varies).

Imagining the avalanche to be a luminescent cylinder (radius r, length x) and defining brightness by the number of photons emitted divided by the projected area, the brightness in the side view will be

$$B_\perp \propto Fn/rx$$
$$\propto E^{-1}[\exp(E^2 t)]/E^{-3/4}\, E^{-1}$$
$$\therefore B_\perp \propto E^{3/4}.$$

And along the avalanche axis

$$B_{//} \propto Fn/r^2$$
$$\propto (E^{-1})[\exp(E^2 t)]/(E^{-3/4})^2$$
$$\therefore B_{//} \propto E^{1/2}$$

The implication for streamers is that they will be brighter if the field is increased and the pulse duration decreased so that the streamer length remains constant.

6.4.2 Photographing streamers

Photographing streamers is not trivial because they emit so little light. They may easily be visible to the eye, but for some purposes it is necessary to arrest their growth at an earlier stage —when they are not. The obvious approach is to collect as much light as possible by using a large camera aperture ($f/2$ or $f/1\cdot5$) and to record with very fast film. One consequence of using large apertures is that the depth of the field is seriously reduced. This is certainly a disadvantage if large chambers are contemplated, and so attention must be paid to the matter. Of course, the practice is to hold the shutter open waiting for a discharge.

Bulos et al. (1967) have discussed quantitatively the factors governing the precision attainable in the photography of streamers. Starting with the lens formula $(1/u) + (1/v) = 1/f$ and the demagnification $M = u/v$, it may be shown that the depth of field in the object space Δu leads to an apparent depth in the image space of $\Delta v = \Delta u / M^2$. The diameter ($d$) of the circle of confusion in the film plane is $\Delta v(D/v)$ where D is the diameter of the lens; $v = f(M + 1)/M$; and the f-number $f_n = f/D$.
Hence

$$d = \Delta v(D/v)$$
$$= (\Delta u/M^2)(D/f)(M/(M + 1))$$
$$= (\Delta u/M^2 f_n)(M/(M + 1))$$

which, for large M, is proportional to $1/M^2$.

Projected back to object space, the diameter of the circle of confusion is

$$Md = (\Delta u/M f_n)(M/(M + 1))$$

The ratio of the diameter of the circle of confusion to the size of a streamer in object space therefore decreases linearly with increasing demagnification.

The accuracy with which a streamer, viewed end on, may be

located may be assumed to be a fraction (1/10, say) of the streamer's apparent diameter (S). This apparent size, after recording, will be given by the square root of the sums of the squares of the following: the actual streamer diameter in space (s); the circle of confusion due to object depth of field; circle of confusion due to image depth of field (i.e. to film jitter); diffraction from the lens aperture; and the resolutions of the lens and the film projected into space.

$$S^2 \approx s^2 + \left(\frac{\Delta u/2}{(M+1)f_n}\right)^2 + \left[\frac{M^2\delta/2}{(M+1)f_n}\right]^2 + [(M+1)\lambda f_n]^2 + \left(\frac{M}{R_1}\right)^2 + \left(\frac{M}{R_2}\right)^2$$

where the focus occurs at the centre of Δu—the extreme depth of field; δ is the extreme jitter in film flatness; λ the wavelength; R_1 and R_2 the resolutions of the lens and the film (both 50% modulation).

In the case of the Stanford chamber (Bulos et al., 1967) $\Delta u = 600$ mm, $M = 70, f_n = 2, \delta = 0.01$ mm, $\lambda = 6.10^{-4}$ mm, $R_1 \sim 50$ lp/mm, $R_2 \sim 70$ lp/mm, and the actual streamer diameter $s \sim 1$ mm.
Applying these values

$$S^2 = (1)^2 + (2\cdot 1)^2 + (0\cdot 2)^2 + (0\cdot 1)^2 + (1\cdot 4)^2 + (1\cdot 0)^2$$

giving an apparent streamer diameter $S = 3$ mm. If the streamer centre be located to $S/10$ the resolution in space becomes ± 0.3 mm.

An optimum demagnification of 70 results if $f_n = 2$ is substituted in the equation for S (minimizing S). An alternative approach is to minimize the signal-to-noise ratio, and this amounts to adjusting the demagnification until the circle of confusion due to the depth of field becomes roughly equal to the projected size of the streamer. This also gives $M \sim 70$.

The fastest film should be employed (a.s.a. >1000). At SLAC, both Kodak 2475 and SO 340 film have been used—with special developing recipes for attaining good definition. Any film requires a threshold intensity before it records: in practice at SLAC f-numbers up to 2 have been satisfactory.

Many practitioners have used the 2475 film. However, the

operating conditions of a chamber often allow a variety of photographic parameters. The author, for example, has used Ilford HP4 (600–1200 a.s.a.) film, $f/2 \cdot 8$, and a demagnification of 20.

Eschstruth et al. (1968) have reported using Kodak SO 340 film which has a speed of 2450 a.s.a. when machine developed in Kodak 19 at 31°C and 55 ft/min. They set the aperture of their 50 mm lens at $f/2 \cdot 2$ so that the effective 60 cm depth of their double-gap chamber was viewed with a diameter of confusion in space of 1·4 mm.

At DESY, the demagnification was 38 with a 35 mm lens (Zeiss Distagon) at a distance of 155 cm. With an f-number set at 2·0, Kodak tri-X aerographic film SO 265 (10,000 a.s.a) was used (Ladage, 1972).

At Frascati (1973), Dr F. Villa reported a new film from Kodak (SO 121) which is eminently suitable for streamer chambers. It has a much finer grain, for better antihalation properties and about the same speed as the SO 265. It has been used with excellent definition at SLAC.

6.4.3 Image intensification

An image intensifier produces a bright optical image of a rather dark scene. When a faintly luminous track in a streamer chamber is focused on to the front of an intensifier, a bright unmagnified image appears at the back and this may be photographed in the conventional manner.

The action of an intensifier is indicated in Fig. 6.4.3.1 (Emberson et al., 1962). An image of the streamer tracks is focused on

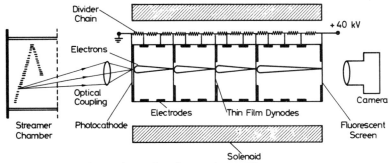

Fig. 6.4.3.1 The action of an image intensifier

the caesium–antimony photocathode. The photons cause the emission of electrons. These are accelerated by the electric field and focused on to the first thin dynode by the magnetic field of the solenoid. The image is thus preserved in the electrons at the first dynode, which consists of a 500 Å film of potassium chloride on a conducting layer of aluminium deposited on a supported 500 Å film of aluminium oxide. At the first dynode, multiplication occurs: typically five secondary electrons may emerge for every 5 keV incident electron, and these will proceed to be accelerated in the second stage where the process is repeated. And so on. At the end of the final stage, the electrons are focused on to a silver-activated zinc sulphide screen where up to 1000 photons may be produced by each electron. Thus with a photocathode efficiency of about 10%, the overall light amplification in a five-stage tube with 40 kV across it might be 10^5 to 10^6. And, of course, the image is preserved on the screen.

Gygi and Schneider (1966) have described the use of a four-stage EMI 9694 image intensifier for the detection of primary avalanches. A constant voltage was applied to the first three stages, and the fourth voltage was pulsed by the trigger system on recognition of a particle in the chamber. By this means, light from noise electrons was minimized: photographic film was not darkened even with a ten-minute exposure. It was observed that a single electron released from the photocathode would be registered on the photographic film. Figure 6.3.3.1 shows photographs taken with the image intensifier: the resolution was 10 line pairs/mm within a diameter of 4 cm.

In a comparison of direct photography and image intensification, Bulos *et al.* (1967) concluded that they gave about the same spatial precision: 0·3 mm in space.

6.5 PERFORMANCE OF STREAMER CHAMBERS

6.5.1 Precision of measurement

Two types of performance must be distinguished. One is the precision with which a particle trajectory may be located in space; the other is the ability to distinguish and count individual streamers. This second property is of relevance to the measurement of ionization and will be discussed later.

Numerical estimates of the spatial resolution attainable with the 20 cm Bedford chamber have been reported by Rice-Evans and Mishra (1969). In an analysis of cosmic-ray tracks, the 35 mm film coordinates of streamers were assessed with a digitizing microscope quoted to have a 2 μm accuracy. Ten points were chosen to identify each track. The results of a statistical treatment are displayed in Table 6.5.1; and it is seen that the uncertainty in the orientation of the 20 cm tracks was 1·5 mrad at its best. As one would expect, the side view of

TABLE 6.5.1

Analysis of measurements on 20 cm cosmic ray tracks. The angular uncertainty is the standard deviation of the orientation of the track. The last column refers to the degree to which the selected measuring points lie off the computed line. (Rice-Evans and Mishra, 1969.)

Streamer regime	Nominal field strength (kV/cm)	Angular uncertainty (mrad)	Standard deviation mm
Merged	12	1·5 ± 0·1	0·34 ± 0·02
End view (projection)	12	2·2 ± 0·24	0·41 ± 0·03
End view (projection)	24	1·5 ± 0·08	0·44 ± 0·02
Side view (isolated streamers)	12	4·2 ± 0·6	0·90 ± 0·07

isolated streamers gave the worst accuracy. The merged streamer regime (at favourable angles) indicated a satisfactory precision. This mode has the advantage of greater luminosity and a stereoscopic track can be recorded with perpendicular views.

The scatter of streamers about a trajectory depends on the drift of ionization from the collision sites and therefore on the time that elapses before applying the high-voltage pulse. Chikovani et al. (1964) estimated that with a delay of 1 μsec, the natural curvature of a 100-cm track due to this scatter would correspond with a maximum measurable momentum of 400 GeV/c—in a field of 10^4 oersted and with an assessment of 140 points.

In their measurements on pion tracks with lengths in the range 12·5–17·5 cm (=225–275 MeV/c), Eschstruth et al. (1968) found that the most probable streamer scatter was 0·35 mm and 0·7 mm for the projected and side views respectively. The delay was 800 ns, and 0·12 mm was believed to be due to multiple scattering. Following the analysis of Gluckstern (1963) they estimated the r.m.s. error in the decay angle ($K_L^0 \to \pi^+\pi^-\pi^0$) in a magnetic field to be 8·4 mrad.

One must beware of two effects that can lead to inaccuracy in locating the tracks. The first is the drift of the ionization electrons in the weak field as the pulse is rising. The second is that streamers do not develop symmetrically about the ionization site; rather they appear centred about a point near the head of the original avalanche. Both effects displace the whole track in the same sense. A good measure of this displacement is obtained by viewing a straight trajectory through both halves of a double-gap chamber where the fields will be in opposite directions. Of course, when the streamers are viewed parallel to the field, these displacements will not be noticed.

6.5.2 Efficiencies and memory times of streamer chambers

The efficiency of a streamer chamber should be 100%. This assumes that the gas mixture is correct and that the high-voltage pulsing arrangement is both satisfactory and reliable. So many ions are liberated along the (say, 20 cm) path of a charged particle, and because each ion has a high probability of initiating an independent streamer, the chance of not producing enough streamers to manifest the track is negligible.

The chamber should also be efficient for any number of simultaneous particles. The author, for example, has seen more than thirty in the Bedford chamber. However, it is conceivable that a vast number of streamers might rapidly reduce the applied field and hence diminish the streamer brightness below the photographic threshold. The streamers exhibit a finite radius and some scatter about the track, and hence may render difficult the distinction of many closely placed tracks.

In the projection mode, the chamber is essentially isotropic; a track should be clearly delineated regardless of its orientation. In the merged streamer regime, good continuous filaments are obtained for angles up to about 30°, after which the quality declines and breaks may appear.

The memory time—i.e. the sensitive time prior to the application of the high-voltage pulse, during which all traversing particles will be duly registered—depends upon the chamber gas. Electrons remain free in noble gases, and therefore in neon/helium chambers the memory times may be 1 ms or more. Impurities emanating from the chamber walls are likely to reduce this value: Eschstruth et al. (1968) found that the density of measurable streamers in their chamber fell by 50% in 30 μsec. A short memory time is advantageous because it reduces the number of undesirable background tracks. Small additions of electron-attaching gases may achieve this object, e.g. 0·1% alcohol was used by Stretz and Perez-Mendez (1969). It is clear that the memory time is essentially the time resolution, i.e. the minimum time interval between events such that two events can be distinguished from one.

Eckardt (1970) has pointed out that tiny traces of the very electronegative SF_6 ($\sigma_{elec.\ cap.} = 1·7 \times 10^{-4}$ cm^2) are best for limiting the memory time, e.g. 10^{-5}% SF_6 results in a 2 μs memory time. This concentration is so small that the SF_6 does not interfere with the streamer mechanism itself.

Continuously to monitor the memory time, Eckardt and Gebauer (1972) circulated the operational gas through an ionization chamber mounted with an alpha-particle ionization source. With a low-frequency alternating potential applied to the electrodes (to exclude the contribution from slow-moving SF_6^- ions), the current registered the free electrons extant and hence the action of the SF_6. This current was used automatically to control a mechanical valve capable of introducing small quantities of SF_6 into the streamer chamber.

In many experiments one would wish to record an event in a chamber only after a logical analysis of pulses from a system of counters has been made and satisfied. This might take up to a millisecond. Under normal operation with an intense beam, hundreds of unwanted tracks could occur in a sensitive time of this order. To overcome this handicap, Eckardt (1970) devised an ingenious method for reducing the memory time to about a microsecond while still retaining the millisecond delay for decision-making.

The technique consists essentially of applying two high-voltage pulses. The first—say, 12 kV/cm—appears within a microsecond of the event. Avalanches are formed. With SF_6 present, the resulting electrons are soon captured. However, in the avalanches a number of metastable atoms will have been created, and because of their long lifetimes, their existence will store information of the track. Penning collisions such as

$$A_{met} + B \rightarrow A + B^+ + e^-$$

or

$$A_{met} + B_{met} \rightarrow A + B^+ + e^-$$

will liberate electrons slowly over a period of milliseconds. (The excitation energy of the metastable states of A must exceed the ionization energy of atoms B or B_{met}.) A second high-voltage pulse (25 kV/cm), applied after the decision-making, will develop photographable streamers from these seed electrons.

The method was successfully tried in the DESY streamer chamber. The first high-voltage pulse, 25 ns, 12 kV/cm was derived from a 250 kV Marx and shaped with a capacitance plus series spark gap. The second pulse, 7 ns, 25 kV/cm, applied milliseconds later, was obtained from another Marx and shaped with a Blumlein. The storage effect worked with a variety of mixtures of Ne, He, Ar, Kr, Xe, Hg and N_2. The best mixture for which excellent tracks were obtained, was the standard 75%/25% composition of neon and helium.

6.5.3 Performance with helium

Some conceivable experiments would benefit if helium were used as a chamber gas. For example, polarization measurements that use the analysing power of helium in nucleon–helium scattering; particle production experiments where the zero-spin, zero-isospin helium nucleus simplifies the interpretation of results; cases where the multiple scattering of low-energy particles must be minimized; and experiments where the primary specific ionization of particles is to be determined (see section 6.6.3).

Working with helium is not as easy as with neon. Rather higher voltages are needed, and the width of the field pulse must be controlled more rigorously if reliable streamers are to be obtained.

Falomkin et al. (1964) produced reasonable tracks in helium at 1 atmosphere, with 150 nsec. 15 kV/cm pulses. At the higher pressures of 4·5 atm and 6 atm, fields of 25 kV/cm and 40 kV/cm were needed to produce streamers with lengths of 3–7 mm and diameters ∼ 0·5 mm. The track brightness hardly changed. The memory time with commercial helium was 100 μsec but with admixtures it could be reduced to a few microseconds. The tracks were photographed at $f/4$. In a later study (1967) these researchers reported tracks consisting largely of luminous streamers whose lengths in the field direction were limited to 1 or 2 mm. They were photographed directly from the side at $f/2 \cdot 8$, and the shunt action of the discharge itself was allowed to shape the pulses. The short lengths were achieved by adding minute quantities of hydrocarbon vapours to the helium, e.g. 0·1% benzene, or hexane, or α-pinene, and even 0·01% of "vacuum rubber vapour".

Other groups have succeeded with helium. Stetz and Perez-Mendez have observed the recoil alpha particles when a helium chamber was bombarded with 80 MeV protons. One difficulty in this type of experiment is that the specific ionization of the alpha particles is so much greater than that of lighter particles. This means that one has to deal with exceedingly faint tracks in the presence of bright trails.

Davidenko et al. (1969) have used helium for measurements on specific ionization. Pictures of finely separated streamers were obtained in 0·6 atm of helium by using an image intensifier. In comparison with neon, the helium streamers were dimmer. There are also fewer of them per cm—corresponding to the differences in specific primary ionization (a factor of 4).

6.5.4 Performance with hydrogen

Great advantages would be gained if hydrogen could be employed in a streamer chamber. A target of protons would then be presented to any beam. At high pressure, i.e. high densities, the streamer chamber would rival the bubble chamber, and also have the advantage of being triggered.

Unfortunately, operation with hydrogen is very difficult. The nobility behave well, but hoi polloi—well, what can one expect? Komarov and Savchenko (1965) have experimented with hydrogen. They used 45 × 45 cm copper-grid electrodes isolated from the 6 cm working gap by perspex plates. They obtained photographable discharges (Fig. 6.5.4.1) by applying 100 ns, 20 kV/cm pulses to the chamber. Both in the side and end (projected) views, a fair indication of the track direction was obtained, and the streamer density averaged to 1·7 per cm. It was

Fig. 6.5.4.1 The best wide-view photograph of a 25 cm track of streamers in hydrogen obtained by Komarov and Savchenko (1965).

observed that the streamer development is rapidly increased as the voltage is raised. Better tracks were obtained by using a 50/50 mixture of hydrogen and helium.

On the other hand, Schmied, Rohrbach and Piuz (1972) tried very high fields of short duration. They constructed three tricoaxial Blumleins incorporating solid araldite dielectric; and were able to apply 500 kV of either 16, 9 or 6 ns to a 9 cm gap chamber—via a 300 kV trigger spark gap. Some initial success was reported; faint trails of streamers in hydrogen at 200 Torr, 33 kV/cm were obtained. They could be photographed both in end-on, and side-view modes with Kodak 2475 and either $f/1·9$ or $f/0·95$ lenses. Nevertheless, a major problem was the prevention of the rapid elongation of the streamers (see also section 9.4.2). It was suggested that because the lifetimes of the significant states in hydrogen were of the order of 0·8 ns, if short streamers were to be obtained it would be necessary to limit the duration of the field also to about a nanosecond.

At these high-voltage gradients difficulty was experienced with breakdown along the chamber walls causing a rapid short-circuit after 6–7 nsec. Curiously, it appears no attempts were made to isolate the electrodes from the sensitive gas with an insulator, although mention is made of choosing the electrode material with a low photoelectric effect to minimize the effect of high-energy photons at the cathode. The conclusion that must be drawn from the work of Rohrbach and his colleagues is that the realization of a hydrogen chamber with a large useful volume for particle experiments will be a very difficult technical feat.

6.6 IONIZATION MEASUREMENT IN STREAMER CHAMBERS

6.6.1 Prospects for measurement

In section 1.2.2 the distinction was drawn between the number of ionizing collisions made by a traversing particle (called primary ionization) and the number of electrons actually liberated (total ionization). It was noted that a measurement of either might give information on the velocity of the particle, or its charge. In section 1.2.3 it was explained that primary ionization is governed by Poisson statistics in which the relative error will be $N^{\frac{1}{2}}$, where N is the number of ionizing collisions along the length of the path. On the other hand, total ionization is described by a Landau-type distribution in which the width, and hence the accuracy of measurement, is hardly improved by lengthening the track. A measurement on the primary ionization would therefore seem preferable.

Initially, the electrons will be located at the collision sites along the particle trajectory. Therefore if the chamber be rapidly triggered one would expect all the electrons at a particular site to participate in the same avalanche. Hence a count of the number of avalanches initiated should give the primary ionization. If a long delay is imposed before triggering, the electrons will diffuse apart and in this case the number of avalanches created might indicate the total specific ionization.

In most chamber regimes, avalanches grow into streamers. The first-developed streamers inhibit neighbouring avalanches, and so the number of streamers is normally less than the number of avalanches created. Nevertheless, the following properties might depend on ionization:

1. Number of streamers per unit length,
2. Number of gaps per unit length,
3. Distribution of gap lengths between adjacent streamers,
4. Streamer brightness,
5. Streamer length along the field.

The first three are similar to measurements made in bubble chambers and nuclear emulsions.

6.6.2 Streamer brightness at Erevan

Asatiani *et al.* (1967) have made a study of the correlation between ionization, streamer lengths and the track brightness when viewed along the field. The principle behind the measurement concerns development times. The total time (t) consists of the time for avalanche multiplication (t_a), plus the time for streamer growth (t_s): ($t = t_a + t_s$). Typically, t_a might be 10–40 nsec; t_s just a few nsec. A small change in t_s yields a big change in brightness, and thus will give a sensitive measure of t_a. And t_a is expected to decrease with higher primary ionization.

In their tests, the Russians used a $50 \times 35 \times 15$ (cm) chamber containing either neon or helium. 150 kV pulses of three fixed durations: 40, 41 and 43 nsec, were applied from a Marx with the aid of a shunt gap. Protons of different known energies were employed, and the brightness of the photographed tracks was measured with a microphotometer. A reasonable linear relationship between brightness and ionization was obtained.

The brightness method however has not become a popular technique. The explanation lies partly in the stringent requirements for the fixed duration pulse.

6.6.3 Primary ionization in Moscow

Davidenko *et al.* (1969) have shown that the specific primary ionization can be measured by counting streamers. The essential point is to apply the pulse before the secondary electrons have had time to drift from the collision sites to initiate their own avalanches. Choosing a criterion that the diffusion length must be less than a streamer radius (0·4 mm), the Russians calculated how the number of streamers would vary as a function of the delay in applying the field. Figure 6.6.3.1 shows that the diffusion in neon is much too fast, whereas in the case of helium

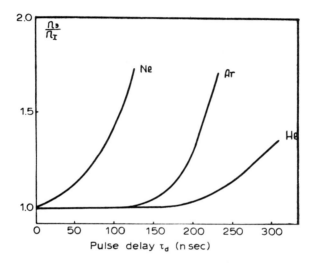

Fig. 6.6.3.1 Calculation of ratio of streamers to primary ionizing collisions as a function of delay in applying the pulsed field. At one atmosphere, the electron diffusion coefficients were taken to be 2000, 370, 310 cm^2 sec^{-1} in neon, argon and helium respectively. (Davidenko *et al.*, 1969)

the number of streamers should indicate the primary ionization if the delay is shorter than 200 nsec.

In their experiments, the streamers were arrested at an early stage so that an image intensifier (gain 1000, resolution 20 lines/mm) had to be used. Measurements confirmed the values for the diffusion coefficients in neon and helium. In both cases, highly purified gases were employed: the gas was in continuous contact with activated carbon, preheated to 100°C at 10^{-2} mm Hg. The r.m.s. deviation (σ) of the streamer centres from the track was measured as a function of the delay (t) and the relation, $\sigma \propto \sqrt{t}$,

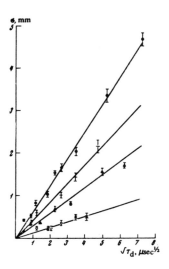

Fig. 6.6.3.2 The rms deviation of streamer centres from the track v. delay. Steepest line: pure neon, 1 atm; next: neon plus epoxy vapour; next: pure helium, 0·6 atm; gentlest: neon, 1 atm, with water or alcohol vapour. (Davidenko et al., 1969)

was observed (Fig. 6.6.3.2). When alcohol, water or epoxy resin vapours were present, the measurement showed the drastic reduction (by factors of 10) in the diffusion that had been noted by Kotenko et al. (1967). This finding has a practical importance: electrons are inhibited from drifting from the collision sites. Photographs of tracks after various delays are shown in Fig. 6.6.3.3.

Counts of the numbers of streamers in a 5 cm chamber filled with 0·4 atm helium, at two delays, gave the distributions shown in Fig. 1.2.1.3. In the case of a 200 nsec delay, expected to correspond to primary collisions, the distribution is essentially Poisson-like; for a 15 μsec delay the distribution approximates to Landau. It should be noted that in the case of a small delay, delta rays occur approximately once every 2 m, and these must be ignored.

Figure 6.6.3.4 shows a measurement of specific primary ionization in 0·6 atm helium, as a function of electron energy. A good indication of the relativistic rise in ionization (see

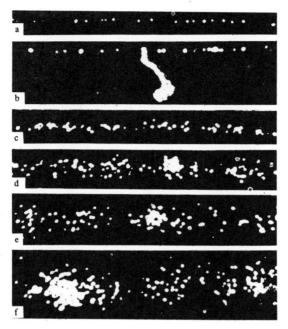

Fig. 6.6.3.3 Minimum ionizing particle tracks in neon at 0·8 atm. The greater the delay (a) 1, (b) 2, (c) 5, (d) 10 μsec, the more numerous and deviant the streamers in the 5 cm tracks. (Davidenko *et al.*, 1969)

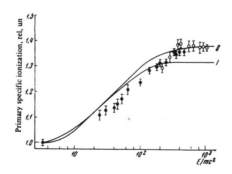

Fig. 6.6.3.4 Relativistic increase of the primary specific ionization measured in a streamer chamber filled with helium at 0·6 atm. (Davidenko *et al.*, 1969.) The solid curves I and II are theoretical calculations from Budini *et al.* (1960) and Kotenko *et al.* (1969)

section 7.5.7) is obtained. Davidenko *et al.* suggest that in a 1·5 m chamber one can determine the ionization to an accuracy of a few per cent, and that this should permit the mass analysis of momentum-selected beams in the region $10 < E/mc^2 < 300$.

6.6.4 Ionization matters at CERN

The avalanche chamber by Gygi and Schneider (1966), discussed in section 6.3.3, manifests primary avalanches with the aid of an image intensifier. Its capacity to measure ionization is indicated in section 9.1.6 where its use as a quark-hunting device is described.

As the frontiers of particle physics expand to higher energies, thought has to be given to the prospects for measurements at these energies. Conventional methods depending upon variations in velocity, such as Cerenkov and time of flight techniques, become more and more difficult to apply as the constant speed of light is approached. A promising line of attack is to rely on the relativistic rise in ionization as a function of particle energy (see section 1.2.1). The rise, which is proportional to $\log(\beta/(1-\beta^2)^{\frac{1}{2}})$ as $\beta \to 1$, has been measured in cloud chambers by Fortune (1960); and it is he (1969) who has pointed to some of the advantages of the streamer chamber for these measurements.

Another application is distinguishing particles that have the same momentum, e.g. identifying pions and electrons in a mixed 200 MeV/c beam. To convey the gist of the technique, the next few paragraphs will present Fortune's treatment. Measurements of the specific primary ionization are concerned and Poisson (normal) statistics will therefore be appropriate.

The relativistic rise (see Fig. 1.2.1.2) in the momentum range between minimum ionization and the Fermi plateau is essentially linear, and the ionization (relative to minimum) may be expressed as

$$i = \alpha \log(p/mc) + \text{const}$$

where α is the slope and m the rest mass. We then have

$$\log m = \log p - i/\alpha + \text{const}$$

which relates the log mass to the momentum and the ionization.

The standard deviation of the logarithm of the particle mass will be

$$s_m = (s_p^2 + (i/\alpha)^2 s_i^2)^{\frac{1}{2}}$$

where s_p and s_i are fractional standard deviations. In practice i/α is about 7, so the mass error is seven times more sensitive to ionization than to momentum error. Errors in momenta need only be a few per cent, so we have

$$s_m = 7 s_i$$

Figure 6.6.4.1 illustrates the prospects for distinguishing pions, kaons and protons for three values of the mass measurement error. For example, to separate protons from kaons, s_i must be so improved to reduce s_m below 20%.

The ionization error is $(n)^{-\frac{1}{2}}$, where n is the number of

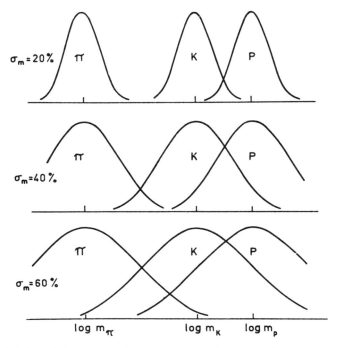

Fig. 6.6.4.1 Mass separation of pions, kaons and protons as a function of precision of mass measurement. (Fortune, 1969)

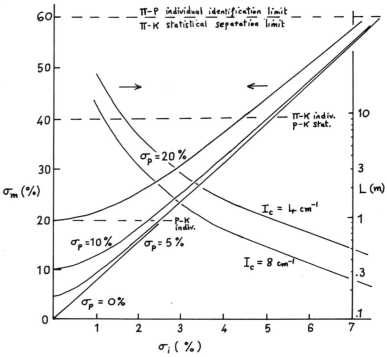

Fig. 6.6.4.2 Dependence of mass error on momentum error, ionization error, ionization density and track length. (Fortune 1969)

countable primary events. (N.B. the Gygi and Schneider one-litre chamber yielded 4 events/cm.) This error will be reduced by lengthening the chamber path. The relationship between the necessary track length L, s_i, s_p, and the resulting error in the mass s_m, for two values of the countable ionization density, is given in Fig. 6.6.4.2. We see that in the region of the linear relativistic rise ($7 < p/mc < 350$), for $I_c = 4$ cm^{-1}, a streamer-chamber length of 1 metre will give individual identification in k–π and p–π mixtures.

Two practical matters handicap the measurement of primary ionization over large track lengths: the resolution of image intensifiers and the production of the pulsed field (20 kV/cm for 4 ns) across a large volume. At the time of writing a capable one-metre chamber appears feasible.

CHAPTER 7

Proportional Chambers

7.1 INTRODUCTION

PROPORTIONAL counters have been used for many years as detectors of charged particles. In their simplest form they consist of a fine wire anode lying along the axis of a cylindrical cathode, the volume being filled with argon plus some quenching

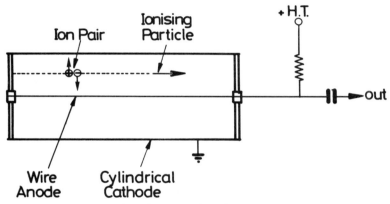

Fig. 7.1.0.1 Coventional proportional counter.

gas (Fig. 7.1.0.1). If the radii of the wire and cathode are r_1 and r_2, respectively, then the electric field at any radius in the counter will be

$$E = \frac{V}{r \ln (r_2/r_1)},$$

where V is the potential difference between the electrodes. The $1/r$ dependence indicates that the field can be very strong near the wire, especially if its radius is small.

When an energetic-charged particle passes through the gas it will create ion pairs along its path, and in doing so will lose about 30 eV every time. The central wire will attract all the electrons produced in the sensitive volume. As the magnitude of the applied voltage is increased, the different regimes depicted in Fig. 7.1.0.2 are encountered. First, there is the "ionization region" in which the charge collected at the anode is just the number of electrons liberated by the traversing ionizing particle. As the voltage is increased, the field becomes stronger and the electrons are able to gain high energies through acceleration. When the proportional threshold is reached, some electrons acquire sufficient energy to cause secondary ionization, i.e. multiplication of the Townsend avalanche variety occurs. The

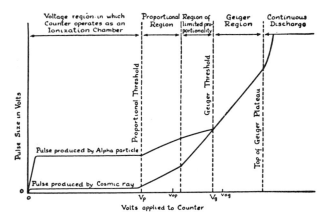

Fig. 7.1.0.2 Operating characteristics of a counter. The α particle creates 10,000 ion pairs in the counter, the cosmic-ray 30. (Korff, 1955)

size of the output pulse still depends on the number of primary ions created. For example, if 40 ion pairs were formed initially, and the multiplication is 10^5, then the output pulse is due to 40×10^5 electrons arriving at the anode.

At still higher voltages the Geiger region is reached, in which all output pulses are of the same size, regardless of the number of primary electrons. In the Geiger regime, the initial avalanche, localized at the wire, initiates a new avalanche in the space to either side. These in turn produce further contiguous avalanches, and the process continues until avalanches have been created all the way down the wire in both directions from the original site. It is thought that the adjacent avalanches are initiated by photo-

PULSE FORMATION IN A COUNTER 247

ionization. The spread of the discharge down the wire may have a measured velocity of 10^7 cm sec^{-1}, and because the mean lives of the excited states are $\sim 10^{-8}$ sec, one can estimate that about ten acts of ionization occur per cm—implying that the mean free path of the pertinent photon is 1 mm (Korff, 1965). This photo-ionization mechanism requires a mixture of two gases—normally a noble gas and an alcohol vapour—so that radiation from excited states of one may ionize the other. The reason that the output pulses are of constant height is now clear; it is because the same discharge occurs along the whole length of wire each time.

From the point of view of high-energy physics, active interest in proportional counters started with the recognition by Charpak *et al.* (1968) that each wire in a multiwire plane might be used as an individual counter. The developments in transistors meant that it was feasible to attach an amplifier to each wire, and as a result a brand new instrument with many advantages has been added to the physicists' toolbox. Known, justly, as Charpak chambers—in honour of their inventor—the devices have been developed in a number of laboratories. However, it has been Charpak and his team at CERN who have subjected the chambers to an extensive and systematic study, as the references in this chapter will clearly show.

7.2 PROPERTIES OF CYLINDRICAL COUNTERS

7.2.1 Pulse formation in a conventional counter

A knowledge of the shape of the voltage pulse obtained at the anode is an important aspect of multiwire proportional chambers, and so first, in this section, we shall derive the pulse produced in a conventional counter. This has been discussed in detail by Wilkinson (1950), and the following is essentially a précis of his treatment.

To begin with, a common fallacy must be exposed. It is wrong to believe that the pulse from the anode of a counter is due merely to the arrival of the electrons; i.e. when an electron arrives, the potential does not instantly change by $-e/C$, where C is the chamber capacity. This picture ignores the effects of charges induced by both the electrons and the positive ions when they lie in the sensitive gas volume.

After the creation of an ion pair, the positive and negative charges will drift in the field, and at time t they will induce charges $-q_+(t)$ and $-q_-(t)$ on the positive collecting electrode. The potential of the anode is then given by

$$P(t) = \frac{q_+(t) + q_-(t)}{C}$$

At the ion formation $t = 0$, we have $q_-(0) = -q_+(0)$, and so $P(0) = 0$. When the electron arrives at the anode ($t = t_1$), all its charge ($-e$) must be induced, so at this moment

$$P(t_1) = \frac{-e + q_+(t_1)}{C}$$

and not just $-e/C$ as the fallacy would have it. Only when the positive ion is collected at the cathode is $q_+(t_2) = 0$ and

$$P(t) = -\frac{e}{C}$$

i.e. the full pulse is obtained only when both ions have been collected.

Expressions for the induced charges depend on the geometry of the counter and may be obtained with the aid of Green's theorem which states that for an isolated system of conductors 1, 2, 3 ... on which charges $q_1, q_2, \ldots$ exist at potentials $P_1, P_2, P_3 \ldots$, if these charges and potentials are replaced by $q'_1, q'_2, \ldots$ and $P'_1, P'_2, \ldots$, we have

$$\sum q_n P'_n = \sum q'_n P_n.$$

Considering an electron ($q_3 = e$) as existing on an indefinitely small electrode (3) between two main electrodes (1, 2), we may take as a first set of potentials $P_1 = P_2 = 0$; $P_3 = ?$ with charges q_1, q_2 and e. Here q_1 and q_2 are the required induced charges. A second set may correspond to the normal operating conditions: P'_1, P'_2, P'_3; $q_1 = q_2 = ?$, $q_3 = 0$. Green's theorem gives

$$q_1 P'_1 + q_2 P'_2 + e P_3 = 0.$$

But

$$q_1 + q_2 + e = 0.$$

Hence

$$q_1 = -e\frac{P_3 - P_2'}{P_1' - P_2'}, \quad q_2 = -e\frac{P_1' - P_3'}{P_1' - P_2'},$$

i.e. the induced charges are determined by the potential at the position of the electron. The same result may be obtained by equating the energy gained by a moving electron to the concomitant loss of electrostatic energy.

It is instructive to consider a parallel-plate ionization chamber (Fig. 7.2.1.1). If an ion pair is created at x_0, the electron will drift swiftly to the anode while the positive ion will appear relatively stationary. From Green's theorem

$$q_+(0) = e\left(\frac{d - x_0}{d}\right)$$

and at time t, if the electron be at x,

$$q_-(t) = -e\left(\frac{d - x}{d}\right)$$

and we have for the anode pulse

$$P(t) = \frac{e}{C}\left\{-\left(1 - \frac{x}{d}\right) + \left(1 - \frac{x_0}{d}\right)\right\}$$
$$= -\frac{e}{C}\frac{ut}{d},$$

where $x_0 - x = ut$, u being the electron drift velocity. Thus the pulse rises linearly until $t_1 = x_0/u$ when we have

$$P(t_1) = -\frac{e}{C}\frac{x_0}{d}.$$

The potential will continue to rise as the positive ions move towards the cathode, until eventually on arrival ($t_2 > 100 t_1$)

$$P(t_2) = -\frac{e}{C}.$$

We notice that in the parallel-plate ionization chamber, the height of the fast rise of the pulse is dependent on the position of the site of primary ionization.

In the coaxial geometry of the conventional cylinder counter,

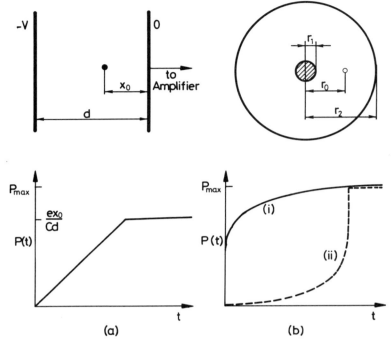

Fig. 7.2.1.1 Typical anode-pulse shapes derived by Wilkinson (1950) for (a) parallel plate ionization chamber; (b) coaxial cylindrical counter: (i) proportional regime (with amplification); (ii) ionization regime (no amplification).

the potential at the site of the ion must be obtained from the field (see Fig. 7.2.1.1)

$$E = \frac{V}{r \ln (r_2/r_1)}.$$

In the case where the counter is operated in the proportional or Geiger regime, the vast majority of secondary ion pairs are created in the high field region, very close to the wire (within two or three wire diameters). We may consider that all the electrons are collected immediately, so that the pulse from one typical secondary ion pair is

$$P(t) = \frac{-e + q_+(t)}{C}.$$

Because of the initial close proximity of the positive ions to the wire $q_+(0) = e$; and so the rise of the anode pulse is almost entirely due to the drift of the positive ions away from the wire. From Green's theorem

$$q_+ = e\frac{P_1' - P_3'}{P_1' - P_2'} = \frac{e}{V}\int_r^{r_2} E\,dr$$

$$= \frac{e}{V}\int_r^{r_2}\frac{V\,dr}{r\ln(r_2/r_1)} = e\frac{\ln(r_2/r)}{\ln(r_2/r_1)}.$$

Now the drift velocity

$$u = \frac{dr}{dt} = KE = \frac{KV}{r\ln(r_2/r_1)}$$

where K is the mobility of the positive ions. Hence

$$\int_{r_1}^r r\,dr = \frac{KV}{\ln(r_2/r_1)}\int_0^t dt$$

and

$$r = \sqrt{\frac{2KVt}{\ln(r_2/r_1)} + r_1^2}.$$

Hence

$$P(t) = \frac{1}{C}\left[-e + \frac{e}{\ln(r_2/r_1)}\ln\frac{r_2}{\sqrt{\frac{2KVt}{\ln(r_2/r_1)} + r_1^2}}\right]$$

$$= -\frac{e}{C}\frac{\ln\left\{\frac{2KVt}{r_1^2\ln(r_2/r_1)} + 1\right\}}{2\ln(r_2/r_1)}.$$

The rise is very fast initially (see Fig. 7.2.1.1) and it continues until the positive ion is collected at time

$$t_2 = \frac{(r_2^2 - r_1^2)\ln(r_2/r_1)}{2VK}$$

when

$$P = -\frac{e}{C}.$$

Finally, for completeness, we shall add the case of the cylindrical chamber without amplification. With the same methods and using $u = K'(E)^{\frac{1}{2}}$ for the electron drift velocity, Wilkinson found the anode pulse shape to be of the form

$$P(t) = -\frac{e}{C} \frac{\ln r_0 - \ln\left\{r_0^{3/2} - \frac{3}{2}K'\left[\frac{V}{\ln(r_2/r_1)}\right]^{\frac{1}{2}} t\right\}^{2/3}}{\ln(r_2/r_1)}$$

for an ion pair created at radius r_0. The pulse shape exhibits a rise which begins slowly but increases rapidly as the electron approaches the wire.

Figure 7.2.1.1 shows the shape of cylindrical counter pulses when operated in the ionization and proportional modes. The very fast rise in the latter case is normally exploited by passing the pulse through a simple *RC* differentiation network before the amplifier. In this manner the length of the output pulse is very reduced without much loss in rise-time or amplitude.

7.2.2 Amplification in a cylindrical geometry

In Chapter 1 the electron multiplication and the growth of avalanches were described in terms of Townsend's first ionization coefficient α, i.e. the amplification in a distance x was given by

$$A = \exp(\alpha x)$$

and it was seen that α could be expressed by

$$\alpha = A'p \exp(-Bp/E),$$

where A' and B were constants, and where E is the uniform electric field.

In the cylindrical proportional counter the field E varies with the radial position, and thus the moving electron experiences a changing α. We therefore write

$$A = N(r_1) = \exp \int_{r_1}^{r_c} \alpha(r)\, dr,$$

where we remember that any electron starting from, say, near the cathode must travel to a sufficiently high field region before ionization can occur. Figure 1.3.5.1 shows how the processes open to an electron vary as it encounters higher fields. In fact, the threshold radius for ionization, r_c, is quite close to the wire (r_1); in a typical counter the mean free path may be about 10^{-3} cm and a multiplication of, say, 1000 only requires 10 such paths. We therefore assume a roughly linear dependence

$$r_c = \frac{r_1 V}{V_c},$$

where V is the applied potential and V_c the ionization threshold potential (see Korff, 1955). In the formula, $N(r_1)$ is the number of electrons that will arrive at the wire as a result of one primary.

An approximate calculation of the amplification has been given by Rose and Korff (1941). In their original paper, a detailed discussion led to the following assumptions. As a result of the cross-sections for ionization (Fig. 1.3.1.3) the electrons capable of ionizing are likely to have energies between the ionization potential and a maximum of about 50 V. In this narrow region, the cross-section may be taken to vary linearly with energy

$$\sigma = a\varepsilon,$$

where a is a constant and ε is the excess of the electron energy above the ionization threshold. Values of a representing the rate of increase of ionization with energy are given in Table 7.2.2 (Rose and Korff, 1941).

Hence we write in terms of averages

$$\alpha = [1/l(\varepsilon)]_{av} = N\sigma = aN\varepsilon_{av},$$

TABLE 7.2.2

Gas	a(10^{17} cm^2/V)
A	1·81
Ne	0·14
He	0·11
H_2	0·46
O_2	0·66
N_2	0·70
NO	0·74
CO	0·83
C_2H_2	1·91
CH_4	1·24

where $l(\varepsilon)$ is a mean free path and N is the number of atoms per unit volume. In addition, the acceptance that the electron energy spectrum decreases monotonically near the wire, i.e.

$$-\frac{\partial \varepsilon}{\partial r} = \frac{CV}{r},$$

where $C/2$ is the capacity per unit length of counter, leads to

$$\varepsilon_{av}(r) = [\phi(0)CV/aNr]^{\frac{1}{2}},$$

where $\phi(0)$ represents the number of slowest electrons and may be normalized to unity. Then

$$\alpha = \left(\frac{aNCV}{r}\right)^{\frac{1}{2}}.$$

Combining equations we have

$$A = \exp Z(aNCr_1 V)^{\frac{1}{2}}[(V/V_c)^{\frac{1}{2}} - 1]$$

which defines the gas amplification in terms of measurable quantities.

Experiments show that the expression may be taken as a rough indication of the amplification. It is most likely to be valid in the low-energy region and, of course, it ignores such effects as photoionization of the gas and the liberation of electrons at the cathode by positive ion or photon bombardment.

In principle, the output pulse from a proportional counter should be directly proportional to the initial ionization. This linearity has been measured by Hanna et al. (1949) to be true to an accuracy of at least 2%. Examples of their spectra are shown in Fig. 7.2.2.1 for differing radiation energies. The linearity was good up to a critical value for the multiplication. This value corresponded to a certain output pulse size, and the cessation of linearity can be explained by the positive ion cloud effectively reducing the field near the wire, and hence also the multiplication.

7.2.3 Fluctuations in pulse height

It would be nice if each output pulse from a proportional counter were to be directly proportional to the energy deposited by the specimen particle. Unfortunately, this is not so because

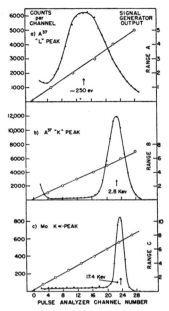

Fig. 7.2.2.1 Proportional counter spectra. The peaks, corresponding to radiations of 250 eV, 2·8 keV and 17·4 keV, show that the percentage width decreases as the energy increases. (Hanna *et al.*, 1949)

of fluctuations in the amplification A, and also in the number of ion pairs produced initially (E/W where W is the average energy required to create an ion pair). These fluctuations will be discussed separately.

A statistical treatment of avalanche growth has been given in section 1.4.7. There it was shown that the probability of an avalanche producing n electrons in a uniform field is given by

$$P(n) = \frac{1}{\bar{n}}\left(1 - \frac{1}{\bar{n}}\right)^{n-1}$$

or for large mean value $\bar{n}$

$$P(n) = \frac{1}{\bar{n}}\exp(-n/\bar{n}).$$

The implication is that there is a very large fluctuation; avalanches with no multiplication are actually e times more likely than those containing $\bar{n}$!! In practice, however, Schlumbohm's (1958) curves showed that the distribution did sharpen at higher fields.

Curran *et al.* (1949) have investigated the distribution of avalanche sizes in a non-uniform field by liberating single electrons from the cathode by a cylindrical proportional counter with ultra-violet light. They found that the distributions were best fitted with the curve

$$P'(n) = n^{\frac{1}{2}} \exp(-n)$$

and the variance was calculated as follows:

$$\bar{n} = \int_0^\infty n^{3/2} \exp(-n)\,dn \bigg/ \int_0^\infty n^{\frac{1}{2}} \exp(-n)\,dn$$
$$= \Gamma(5/2)/\Gamma(3/2) = 3/2$$
$$\overline{n^2} = \int_0^\infty n^{5/2} \exp(-n)\,dn \bigg/ \int_0^\infty n^{\frac{1}{2}} \exp(-n)\,dn$$
$$= \Gamma(7/2)/\Gamma(3/2) = 15/4$$
$$\therefore \quad \sigma_n^2 = \overline{n^2} - (\bar{n})^2 = 3/2$$

and

$$\frac{\overline{n^2} - (\bar{n})^2}{(\bar{n})^2} = \frac{2}{3} = 0.666.$$

Their experimental distribution gave a value of 0·696, so they chose an average value of 0·68 for the variance of the multiplication process (Fig. 7.2.3.1).

It is to be noted that there is a contradiction between the two distributions [$P(n)$ and $P'(n)$]. More recently, however, J. Byrne has justified the Curran formula theoretically (see Curran and Wilson, 1965).

The fluctuations in the number of ion pairs created in a volume of gas ($m = E/W$) have been analysed by Fano (1947). Taking the case in which the energy dissipated is fixed (E_0), thus implying that the expected number of ions is $\bar{m} = E_0/W$, Fano shows that the fluctuation in m might be expected to lie in the range

$$\frac{1}{2\bar{m}} > \frac{\overline{m^2} - (\bar{m})^2}{(\bar{m})^2} > \frac{1}{3\bar{m}}.$$

Thus the expected fluctuation is narrower than if the number of ionizations were governed by the Poisson distribution, in which case

$$\frac{\overline{m^2} - (\overline{m})^2}{(\overline{m})^2} = 1.$$

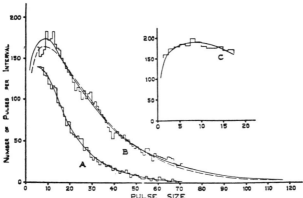

Fig. 7.2.3.1 Pulse-size distribution for single electrons in a proportional counter. The broken curve represents a fit of the form $n^{\frac{1}{2}} \exp(-n)$. Curve A corresponds to 3280 V, B and C to 3550 V. (Curran et al., 1949)

O. R. Frisch (1948) has shown theoretically how to combine the fluctuations due to the above two effects, and gives for the variance of the output pulse the relation

$$V_p = V_m(\bar{n})^2 + V_n\bar{m},$$

i.e.

$$\frac{V_p}{(\bar{n}\bar{m})^2} = \frac{V_m}{(\bar{m})^2} + \frac{V_n}{(\bar{n})^2\bar{m}},$$

where again V_m and $\bar{m}$ are the variance and mean of the number of ions initially liberated in the counter, and V_n and $\bar{n}$ are the corresponding quantities in the multiplication.

West (1953) has compared experimental results with Fano's estimates. Assuming a mean value of 25 eV per ion pair to convert from the mean number of ions ($\bar{m}$) to the energy of the radiation, he found that $V_m = \bar{m}/3$ fitted the data best, especially at low energies (< 10 keV). Combining the values of West and Curran, we have

$$\frac{V_p}{(\bar{n}\bar{m})} = \frac{1}{3\bar{m}} + \frac{0\cdot 68}{\bar{m}} = \frac{1}{\bar{m}},$$

from which it is possible to conclude that the additional spread introduced by the multiplication process is not serious. West (1953) has also discussed other possible sources of spread such as negative-ion formation (significant if impurities are present),

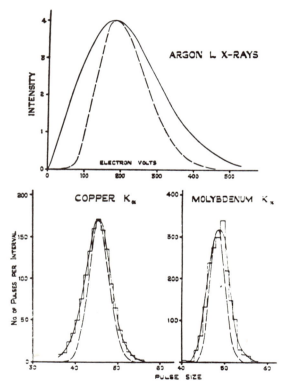

Fig. 7.2.3.2 Analysis of line-width for monokinetic radiations. The outer peaks are the experimental distributions. The inner peaks represent the contribution to the width by the gas multiplication process, the remainder being explained by the fluctuation in the initial ionization. (Curran *et al.*, 1949)

THE ELECTRIC FIELD

variations in voltage, and variations in the position and diameter of the wire (normally small).

It will be remembered that the distribution $n^{\frac{1}{2}} \exp(-n)$ fitted closely the result of Curran *et al.* for avalanches produced by single electrons. This function may therefore be applied to typical X-ray spectra, in which many primary ion pairs are created, to ascertain the relative contributions to the spectrum width by the fluctuations in ionization and multiplication. In Fig. 7.2.3.2 it is seen that the contributions are of the same order.

7.3 PROPERTIES OF MULTIWIRE CHAMBERS

7.3.1 The electric field

In the Charpak chamber, with its wire plane lying between two outer plane electrodes, the equipotentials are as shown in Fig. 7.3.1.1. They were obtained with the conductive paper analogue procedure. Although this is not an accurate method, it is clear that the outer potentials are practically planar, but in the neighbourhood of the wire they circle concentrically—as in a cylindrical counter.

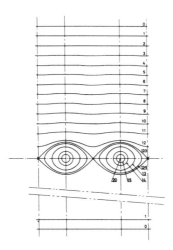

Fig. 7.3.1.1 Equipotentials in a chamber comprised of a central wire-plane anode placed between flat external cathodes. The wire diameter is 4×10^{-3} cm, the wire spacing 0·3 cm and the cathode separation 1·5 cm. (Charpak *et al.*, 1968)

G. A. Erskine (1970) has made some numerical computations of the fields, potentials and wire charges for the standard configuration and also for the special cases of a displaced wire and a variation in wire diameter. He used a rather precise expression based on the complex potential, which was valid for thin wires and which catered for truly plane outer equipotentials.

An earlier expression used by Erskine (see Charpak, Rahm and Steiner, 1970) for the field in space outside an infinite grid of equally spaced infinitely thin wires (see Morse and Feshbach, 1953) was

$$V = q \ln \left[\sin^2 \left(\frac{\pi x}{s} \right) + \sinh^2 \left(\frac{\pi y}{s} \right) \right].$$

Here s is the wire spacing, q the charge per unit length on each wire, and the coordinates x and y relate to an origin centred on one wire; x is parallel to the wire plane, y perpendicular (Fig. 7.3.1.2).

Along the three lines of symmetry $x = 0$, $y = 0$ and $x = s/2$, the potentials may be calculated from

$$V(0, y) = 2q \ln \sinh \frac{\pi y}{s} \rightarrow Ey = \frac{2q\pi}{s} \coth \frac{\pi y}{s}$$

$$V(x, 0) = 2q \ln \sin \frac{\pi x}{s} \rightarrow Ex = \frac{2q\pi}{s} \cot \frac{\pi x}{s}$$

$$V(s/2, y) = 2q \ln \cosh \frac{\pi y}{s} \rightarrow Ey = \frac{2q\pi}{s} \tanh \frac{\pi y}{s}.$$

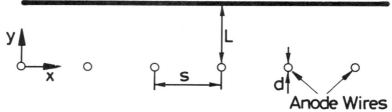

Fig. 7.3.1.2 Dimensions in a multiwire chamber.

Although these formulae are valid only for infinitely thin wires, Charpak et al. (1970) have calculated that for a wire of 100 μm thickness and 1 mm spacing, $\sin(\pi y/s)$ and $\sinh(\pi y/s)$ differ by only 8×10^{-3}, and the change in the field distribution is negligible.

The field becomes uniform very quickly as one moves away from the wire plane. At $y = s/2$ the difference between $\tanh(\pi y/s)$ and $\coth(\pi y/s)$ is 9%, and at $y = 1\cdot 25s$ the difference is only 0·1%.

In the practical case where $L \gg s \gg d$ both of Erskine's expressions yield an approximate formula which enables one to calculate the charge on the wire from the applied potential V_0 and the dimensions:

$$q = \frac{V_0}{2\left[\left(\frac{\pi L}{s}\right) - \ln\left(\frac{\pi d}{s}\right)\right]}.$$

In a typical chamber, $s = 2$ mm and $L = 8$ mm, with an applied voltage of 4000 V, the field at the wire is $2\cdot 2 \times 10^5$ V/cm.

The field at the wire depends on the stored charge which depends on the wire spacing. Table 7.3.1 shows how the wire capacitance per unit length ($C = q/V_0$) varies with the spacing. It must be noted, however, that this is the capacitance concerned in charging the wire to a high voltage. The capacitance relevant to signal production is the capacitance of the wire to earth, which is considerably higher, e.g. 20 pF m^{-1} for $s = 2$ mm.

TABLE 7.3.1

Capacitance of a wire in the central plane with respect to the outer electrodes for $L = 7$ mm
(Charpak, Rahm and Steiner, 1970)

Wire diameter	Wire capacitance (C) (pF m^{-1}) (for given spacings)		
	$s = 3$ mm	$s = 2$ mm	$s = 1$ mm
20 μ	4·97	3·85	2·25
30 μ	5·21	3·96	2·28
40 μ	5·30	4·04	2·31
50 μ	5·41	4·11	2·34

To maintain the same charge on 20 μm wires, i.e. to ensure the same amplification, the table shows that the voltages have to be increased in the ratios 1, 1·3, 2·2 for 3, 2, 1 mm spacings, respectively. Smaller wire spacings will therefore require higher voltages, which may entail breakdown difficulties. The overall voltage may be reduced by decreasing the gap L to, say, 3 mm, but this is at the cost of raising the capacitances to 7·94, 6·81 and 4·56 pF.

7.3.2 Amplification and pulse formation

In the multiwire chamber, the electron amplification occurs in the high-field region near the wires. As we have seen, the field in this region is radial and is given by $2q/r$, which means that many of the properties of the conventional cylindrical counter will apply also to the Charpak chamber.

Electrons liberated by ionization in the main volume of the device will drift steadily in the uniform field towards the wire plane. On approaching a wire the increasing field will impart sufficient energy to the electrons for secondary ionizing collisions to occur, and an avalanche will develop. The expression predicted by Rose and Korff (1941) for the amplification may be written (see section 7.2.2):

$$A = \exp \sqrt{(2aN)} f(r_1) \sqrt{V} [(V/V_c)^{\frac{1}{2}} - 1],$$

where V is the applied potential, V_c the ionization threshold potential, and where the capacitative term $f(r_1)$ can, in view of Erskine's (1970) calculations be written

$$f(r_1) = 1/[\pi L/s - \ln(2\pi r_1/s)]^{\frac{1}{2}}.$$

Here r_1 ($=d/2$) is the wire radius as before. The observed amplification in multiwire chambers agrees fairly well with this formula over a wide range, but at higher voltages it begins to fall below the predicted value.

In section 7.2.1 it was shown that the formation of a pulse at a wire operating in the proportional regime is due not to the arrival of the electrons at the wire, but to the changing charge induced on the wire by the positive ions as they drift away. Nevertheless, the pulse rises rapidly. Rahm (1970) has shown that Wilkinson's expression for the growth of the pulse may be extended to multiwire chambers. The new expression for N positive ions leaving the site of the anode wire is

$$Q(t) = \frac{q}{V} Ne \left[\ln \{e^{2\beta(t+t_0)} - 1\} - \ln (e^{2\beta t_0} - 1) \right],$$

where q is the charge per unit length of wire, K the mobility of the ions ($u = KE$), and

$$\beta = \frac{2q\pi^2 K}{s^2}, \quad t_0 = \frac{1}{\beta} \ln \cosh \frac{\pi d}{2s}.$$

Charpak (1970) has taken the example of a chamber with dimensions $s = 2$ mm, $L = 8$ mm. With an applied voltage of 3000 V and assuming an ion mobility in argon of $1 \cdot 3$ cm s^{-1} V^{-1} cm^{-1}, he calculates the values

$t_{max} = 145$ μsec; $t_0 = 1 \cdot 8$ nsec and $\beta = 7$ μsec.

In multiwire chambers one is concerned with times certainly less than 7 μsec, in which case the expression assumes the Wilkinson form

$$Q(t) = \frac{qNe}{V_0} \ln \frac{t + t_0}{t_0}.$$

The calculations give for the rise of the pulse:

t	Q/Q_{max}
1 nsec	0·015
2 nsec	0·026
10 nsec	0·065
50 nsec	0·120
100 nsec	0·140
500 nsec	0·180
1 μsec	0·200
10 μsec	0·320
100 μsec	0·775
145 μsec	1

To recall, the approximation has been made that all the ions were created at the wire radius, say, 10 μ. In the example, the 145 μsec is the time for these ions to reach the outer electrode. The long slow rise to its final value corresponds to the ions drifting outwardly in the uniform field. The significant fact is, however, that 6·5% of the signal is collected in 10 nsec. It is this very fast initial rise that must be exploited if timing precision is required. It is done with an RC differentiation circuit to produce the pulse of Fig. 7.3.2.1.

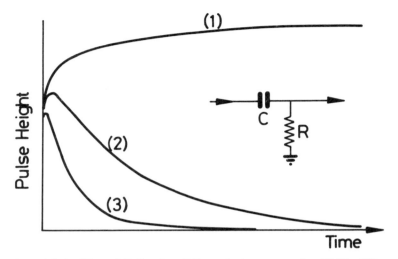

Fig. 7.3.2.1 Use of RC pulse-differentiation network: (1) Undifferentiated pulse from a proportional counter; (2) differentiated pulse; (3) differentiated pulse with a smaller time constant. ($= RC$)

The typical situation is that an energetic-charged particle passes right through the chamber; in which case the electrons liberated in the primary ionization will have varying distances to drift before encountering the high-field region and producing their avalanche. The time of arrival of the avalanche will thus vary between 0 and 200 nsec. As a result, the anode output pulse comprises many small pulses whose commencement is spread over 200 nsec. The final pulse can thus be expected to rise in an approximately linear fashion.

We have seen that a negative pulse is obtained on a wire due to the drift of positive ions away from the wire. Complementary to this will be the similar but opposite (i.e. positive) pulse produced on the cathode. In addition, there will be positive pulses produced on anode wires adjacent to the wire on which the avalanche concentrates. This too is a consequence of the positive ion drift. The negative-going front on the active wire will induce a positive-going front on the adjacent wires. Typical multiplications of the order of 10^4 or 10^5, give signals, after differentiation, of several millivolts when developed in a load of 15 pF in parallel with 10 kΩ. Higher amplifications are possible but good linearity is lost. Charpak et al. (1970) calculate that a wire might be charged to a standing value of $1\cdot2 \times 10^{-11}$ coulombs/mm. This is to be compared with the charge contained in a cloud of, say, 5×10^6 ions—$0\cdot8 \times 10^{-12}$ coulomb. These values are consistent with the hypothesis of Hanna et al. (1949) that the ionic space charge so reduces the field at the wire that the amplification is diminished.

7.4 OPERATION, CONSTRUCTION, AND RELATED PROBLEMS

7.4.1 The Charpak prototype proportional chamber

The main object of any design is to place a plane consisting of extremely fine parallel wires, each, say, 2 mm away from its neighbour, symmetrically between two outer plane electrodes. The system must be supported in a rigid gas-tight box that will maintain a satisfactory insulation between the electrodes.

Details of the prototype chamber by Charpak, Bouclier, Bressani, Favier and Zupančić (1968) are presented in Fig. 7.4.1.1. The central wires, of 0·04 mm diameter stainless steel, were stretched on an araldite frame between outer electrodes comprised of stainless-steel mesh (made from 0·05 mm diameter wires, 0·5 mm apart). The central wires were 7·5 mm away from the mesh. Around the walls a guard strip at earth potential (the anode potential) prevented breakdown to the wires along the surface of the dielectric. The two extreme wires in the central plane were thicker than the others to avoid too high a field gradient. Each wire was connected to an amplifier with an input impedance of 10 kΩ and commonly of a simple three transistor design with a gain of 200.

The chamber, whose volume was defined with thin (80 μm) mylar windows, was flushed at atmospheric pressure with a flow of ordinary argon bubbling through an organic liquid at 0°C. Ethyl alcohol, n-pentane, and heptane were all tried. With

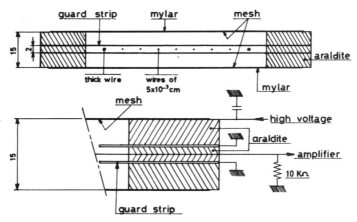

Fig. 7.4.1.1 Details of the construction of the prototype multiwire proportional chamber. (Charpak *et al.*, 1968)

a steady negative voltage applied to the mesh electrodes, a 3 mm wire spacing, and an argon–alcohol gas mixture, the threshold for amplification was 1100 V and the proportional region (in which the Rose–Korff relation holds) lay between 1400 V and 1700 V. At 1400 V the amplification was 50. It was

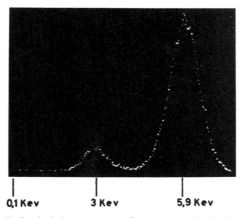

Fig. 7.4.1.2 Pulse height spectrum from one wire in the proportional chamber when irradiated with 5·9 keV X-rays from ^{55}Fe. The small peak is the escape line at 3·0 keV. The wire-spacing was 3 mm, the voltage 3750 V and the gas argon-pentane. (Charpak *et al.*, 1968)

possible to work beyond 1700 V at a cost of losing strict linearity. At 1800 V the amplification rose to 5000, at which value an output pulse of 5·5 mV could be taken directly from the wire, on a 30 pF load. A higher gain was acquired with n-pentane, or heptane, mixed with ordinary argon. For example, 100 mV pulses could be obtained directly from a wire with n-pentane, without entering the Geiger region.

In Fig. 7.4.1.2 is shown the pulse-height spectrum from one wire when the chamber was irradiated with 5·9 keV X-rays from ^{55}Fe. Two peaks are clearly resolved. The main peak corresponds to cases where all the energy has been absorbed; the lower energy peak to the energy deposited by the photoelectron when the corresponding 2·9 keV X-ray escapes. With a high-quality amplifier the resolution was found to be 15%.

7.4.2 Large chambers, electrostatic problems

When large chambers are contemplated, a number of conflicting requirements present themselves. Fully to exploit the $1/r$ field, very fine wires are preferable. On the other hand, the wires must be strong enough to withstand the tension necessary to hold them taut and planar. At CERN, gold-plated molybdenum or tungsten wires of 20 μm diameter are commonly used. The maximum allowable tension for molybdenum of this size is 10–15 g.

Erskine (see section 7.3.1) has extended his calculations to determine what happens if either a single wire is displaced by a small amount or one wire has a different diameter. In both cases the charge q stored in the wire will be affected. Charpak (1970) has reported that the effect of displacing one wire by (only) 0·01 mm in the plane of the wires, is that the charge on the displaced wire is altered by about 0·2% and that on its adjacent wires by 1·6% [sic]. Remembering that the gain varies exponentially with the charge, the surprising consequence is that although the wire itself is little affected, the difference in amplification between the neighbouring wires is as much as 25%! Movements out of the plane do not alter the multiplication nearly so much. Anyhow, it is clear that care has to be taken in the manufacture of wire planes. At a large establishment like CERN, considerable ingenuity has gone into the devising of wire-winding machinery (Alleyn et al., 1968).

Erskine's calculations also gave an approximate formula for the change of charge accompanying a change in the diameter of one wire:

$$\frac{\Delta q}{q} = \frac{1}{\ln(\pi d/8L)} \frac{\Delta d}{d}.$$

A 10% change in diameter causes a 25% difference in amplification in argon/CO_2, and 12% in 75% argon/25% isobutane.

Another problem which arises when large chambers are to be built is caused by electrostatic forces. All the anode wires will have an excess of positive charge; they will repel each other and tend to assume the pattern of Fig. 7.4.2.1. (Another possible contribution—the anode–cathode attraction—may be ignored if the wire spacing is small compared with the size of the gap.)

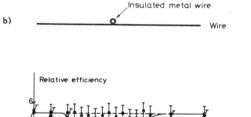

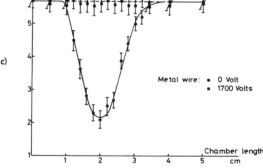

Fig. 7.4.2.1 (a) Displacement of long anode wires due to mutual electrostatic repulsion. (b) Retaining the anode wires in a plane by supporting them with wires sheathed with a poor insulator. (c) Shows that a restoring potential is needed in the support wires to obtain satisfactory efficiency. (Bouclier et al., 1970)

Trippe (1969) has studied this repulsive effect, and it turns out that for wires of length l, the tension (in dynes) is given by

$$T \geqslant \frac{V^2 l^2}{4\pi^2 s^2},$$

where V is the applied voltage (in e.s.u.). The tension is about 100 g for 3000 V if $s = 2$ mm, and this is twice the strength of 20 μm molybdenum wire! The maximum practical length for 20 μm wires is about 25 cm.

Fortunately, greater lengths are feasible if the wires be supported. A number of means to constrain the wires within the median plane have been tried. The first (Dieperink et al., 1969) was to glue taut nylon lines at 15 cm intervals orthogonally to the wires. This proved not to be ideal as the particle-detection efficiency declined over a distance of ~ 2 cm along the wire in the neighbourhood of the lines. The reason for the decline is not established, but it may concern the build-up of charge on the surface of the nylon. Kleinknecht et al. (1969) have found that a 20 μm nylon thread interwoven every 20 cm leaves an inefficient length of only 0·5 mm. Bouclier et al. (1969), on the other hand, have been able to restore the efficiency to its original level by supporting the anode wires with vinyl-coated wires (Fig. 7.4.2.1). These insulated wires were raised to a voltage, typically ~ 1200 V, which overcame the field effect. It was found to be imperative to use a relatively low resistivity insulator (vinyl) rather than a good insulator (teflon). Presumably the former allows the deposited charge to leak away.

7.4.3 Gas mixtures

The main aim in the selection of suitable gas mixtures for Charpak chambers is a high multiplication factor for a given voltage. Large output pulses will require less subsequent electronic amplification. High multiplication will allow lower-applied voltages and hence reduce the breakdown problems associated with high-static voltages.

It has already been mentioned that in the early studies by Charpak et al. (1968) argon at atmospheric pressure was used as the main counter gas with an admixture of either ethyl alcohol, n-pentane or heptane. Argon, being a noble gas, has zero electron affinity and it will therefore allow a high degree of multiplication. It is also cheap. The organic vapour will quench the discharge, i.e. it will neutralize the positive ions and prevent electrons being ejected from the cathode.

Later studies by Bouclier et al. (1970) have indicated that a good performance may be obtained with either isobutane, or carbon dioxide, mixed with the argon. Isobutane has the advantage of a high stopping power for energetic particles, as the table of energy losses shows.

TABLE 7.4.3 (from Bouclier et al., 1970)

Gas	Energy loss in keV cm^{-1} at N.T.P. for 1·3 MeV electrons
H	0·34
He	0·32
Ne	1·44
A	2·50
Kr	4·9
Xe	10·5
CO_2	3·3
CH_4	1·5
Isobutane	5·3

The studies included measurements on the amplification, and the threshold and operational voltages for gas mixtures in various proportions (Fig. 7.4.3.1). With a 2 mm wire spacing and a cathode–anode distance of 7 mm, voltages of up to 7000 V were applied. The larger the proportions of CO_2 or isobutane (→60%) the higher the applied voltage had to be. Pulses of up to 400 mV were obtained from the wires and it was estimated that the output range 0·5 mV–200 mV corresponded to a range of amplification $10^3 \leqslant A \leqslant 10^6$.

A "magic gas" that gives an extraordinarily large amplification has been found by Bouclier et al. (1970). In the course of their investigations of the characteristics of various mixtures of argon, helium CO_2, freon-12 (CCl_2F_2), freon-13 ($CClF_3$), freon-13 Bl (CF_3Br), and freon-12 Bl ($CClF_2Br$), they found that for a particular mixture [argon + isobutane (16%) + freon-13 Bl (<0·3%)] multiplications of the order of 10^8 were obtained. This is a factor of 100 greater than is customary and corresponds to an output pulse of about one volt on an infinite load. A remarkable observation was that even the smallest pulses (initiated by one ion pair) were about 50 mV.

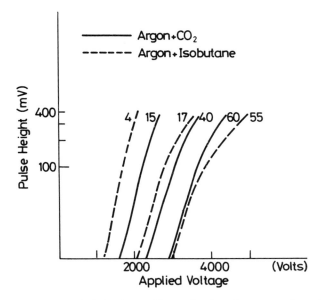

Fig. 7.4.3.1 Pulse height in a Charpak chamber as a function of applied voltage and gas mixture; for 5·9 keV X-rays, with a load of 100 kΩ, 35 pF. The numbers represent the respective percentage of CO_2 or isobutane. (Bouclier et al., 1970)

This enhanced amplification in a particular mixture is not properly understood. It is thought that normally, in a conventional proportional counter, an avalanche is limited (i.e. becomes non-linear) when the positive ion-space charge reduces the field at the wire. Bouclier et al. speculate that negative freon ions are produced in such a fashion as to reduce the effect of the positive ions and allow the multiplication to continue. It appears that the isobutane is necessary for the production of the peculiar space charge and, with the freon, for quenching the photons sufficiently to prevent Geiger action.

Magic gas now enjoys widespread use, partly because

hybrid-integrated circuits with a sensitivity of 5 mV may then be employed. The magic recipe recommended by Charpak, Fischer et al. is (by volume):

$$0.5\% \text{ freon-13 B1} \quad (CF_3Br)$$
$$24.5\% \text{ isobutane} \quad (\text{iso } C_4H_{10})$$
$$75\% \text{ argon}$$

It has been observed that the performance of counters containing a portion of isobutane deteriorates rapidly. For example, after an irradiation of 10^7 beta particles/cm^2, the plateau with the standard magic gas is much reduced and the voltage at which a self-sustained discharge commences is lowered. It appears that the isobutane polymerizes, and the secondary products contaminate the electrodes. Charpak, Fischer et al. have advocated the introduction of an additional quenching agent which is non-polymerizing and has an ionization potential below that of isobutane (10·6 eV). In this event, an exchange process will take place, and the positive ions reaching the cathode will be only those of the harmless additive. Satisfactory results were obtained with propylic alcohol (or 1-propanol, C_3H_7OH, i.p. 10 eV) and methylal (($OCH_3)_2CH_2$, i.p. 9·7 eV). An addition of 4% of the latter led to no change in the amplification properties of the magic gas, and yet the performance remained good even after an irradiation of 5×10^{10}/cm^2. (See Fischer et al., Frascati report, 1973.)

Grunberg et al. (1970) have added electronegative gases to argon to reduce the sensitive region of a multiwire chamber to narrow cylinders around the wires. In any gas, avalanches will grow as long as the mean free path for electron attachment is larger than the mean free path for ionization. Outside the cylinders, the electrons will be attached to form heavy negative ions which cannot participate in the multiplication. They chose ethyl bromide (bromine is very electronegative) and were able to reduce the cylindrical regions to diameters of 3 mm for wire spacings of 5 mm. Inside this diameter, of course, the field is strong enough to ensure a short mean path for ionization. This technique is especially valuable when tracks that are inclined to the plane of the wires are to be handled. Instead of pulses on, say, five wires, only one gives an output. Satisfactory results were obtained for 10%, 20% and 60% proportions of ethyl bromide. Similar behaviour may be obtained with other electronegative gases, e.g. freon-12 (see Fig. 7.5.1.2).

This technique of reducing the sensitive regions around the wires by adding halogenated components may be extended radically to improve the performance. For example, Merkel (1971) using CF_3Br in an arrangement of four successive staggered chambers was able to diminish the time resolution to a few nanoseconds and the spatial precision (σ) to 0·15 mm for a wire spacing of 1 mm.

7.4.4 Constructional details

Many designs for multiwire proportional counters have been similar to the prototype arrangement of Charpak et al. (1968) (section 7.4.1). For example, Koester (1970) has used 6·4 mm thick frames made of epoxy fibre glass milled flat to 50 μm to provide the anode–cathode spacing. When he tried methyl methacrylate he observed spurious pulses which might be explained by the accumulation of charges on the highly insulating surface. The negative electrodes were made of copper-coated mylar foil (25 μm Cu on 38 μm mylar) or aluminium foils coated with acquadag. Uncoated aluminium showed a sensitivity to light which produced intolerable noise. The 50 μm anode wires, creating a sensitive area of 20 × 20 cm^2, were set at 3·2 mm spacings by 50 μm grooves milled in the frames. Guard strips and thick boundary wires were again used.

Gold-plated molybdenum wires have been popular. Bemporad et al. (1970) used such wires with a 30 μm diameter to determine the spatial coordinates of 16 GeV/c particles in a beam. An alternative construction has been tested by Neumann and Nunamaker (1970), in which the anode wires are supported by a slightly conductive (resistivity 10^{12} Ω cm) film. A highly insulating film would allow charges to build up and distort the fields around the anode wires. If desired the wires may be glued to the film with an epoxy cement of the same resistivity.

Large chambers have been fabricated by many groups, e.g. Schilly et al. (1971) and Charpak, Fischer, Minten et al. (1971). In the Charpak design, each complete (40 × 150 cm^2) chamber contained three independent gaps with sense wires stretched horizontally, vertically and at 15° to the vertical. The seven demountable frames were of extruded fibre-glass epoxy compound. Printed circuit boards for high-voltage and sense wire planes were glued to the frames. The frame assembly had to be supplied with aluminium clamps to counteract the effect of 50 gm tension in each 20 μm thick gold-plated tungsten sense wire. The wire spacing was 2 mm and the tension was controlled

Fig. 7.4.4.1 A multiwire proportional chamber with Professor Charpak to the left. (*Photocern*)

to 2%. Copper–beryllium wires, 100 μm thick, spaced at 1 mm, tensioned to 140 gm, comprised the high-voltage electrodes. The sense wires were held in place with the aid of orthogonal insulated wires as explained in section 7.4.2.

Operational hints mentioned by the CERN group included cleaning: by brushing the wires in turn with amylacetate, alcohol and liquid freon. Considerable noise is often observed for a day or two with new chambers, but this may be rapidly removed by inverting the high-voltage polarity for a few seconds (set at 1000 V below normal plateau).

7.5 PERFORMANCE OF MULTIWIRE CHAMBERS

In this section we shall discuss the characteristics of multiwire chambers that will determine their usefulness as particle detectors. Paramount among these are the efficiencies of practically 100%; the ability to handle any number of simultaneous pulses; the localization in space to 2 mm and below; the fast response and the time resolution of 0·4 μsec and below; and satisfactory performance in a magnetic field.

7.5.1 Particle efficiency

The efficiency of a chamber for recording particles will depend on the probability of ion pairs being produced within the gas. This will be a function of the path length in the chamber and the specific ionizations. In the absence of an electronegative gas all the liberated electrons would drift along the field lines to the nearest anode wire where they would produce the avalanche. This drifting is relatively slow and quite some time must elapse before pulses are initiated by electrons released near the cathode. In practice, one may not wish to wait; the time is often limited electronically so that an effective collection distance is defined. If the wire spacing be 2 mm, this effective path length might be chosen to be 2 mm, so that the efficiency will be approximately constant for all particle approach angles.

Sometimes, one may wish to impose a threshold so that the electronics will respond only to events that liberate a minimum number of electrons. The efficiency is then going to be related to the number of delta rays created in the selected region which yield this minimum. The efficiency for particle detection will be given by $1 - e^{sx}$ where x is the path length and s the specific ionization for the appropriate delta rays. Assuming a Landau distribution (see section 1.2.3) Charpak (1970) has calculated, for a 2 mm region in argon, how the efficiencies vary with the threshold. With the minimum set at one ion pair the theoretical efficiency is 99·7%, for ten it is 99·2% and for fifteen 96·4%.

In practice, the chambers give an efficiency of practically 100%. Figures of 99, 99·5 and 99·9% are commonly quoted—for example, Bemporad *et al.* (1970) state better than 99% both for simple and many simultaneous tracks.

Each liberated electron will travel to an anode, but those created in a space equidistant between two anode wires may take longer to arrive. When many ions are formed in this space an avalanche on both wires is likely. This may be seen in the efficiency curve of Koester (1970) which was obtained by moving an orthogonal beam across the wire plane and observing the manner in which the wires responded. The efficiency for observing a pulse on one wire or the other was found to be greater than 99% (Fig. 7.5.1.1).

The efficiency will depend on whether electronegative gases are present to attach some of the electrons to form heavy negative ions, and thus prevent them forming avalanches. They may be impurities (e.g. oxygen in air) or they may be added purposely. Grunberg *et al.* (1970), by adding quantities of ethyl

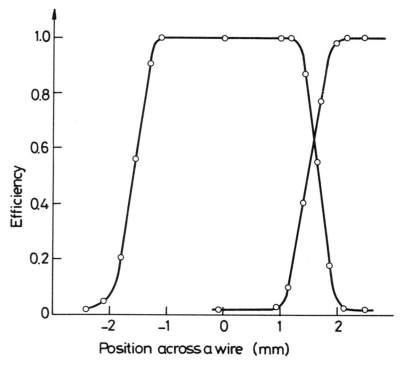

Fig. 7.5.1.1 Individual wire efficiencies versus beam position. (Koester, 1970)

bromide, have been able to restrict the region of high efficiency to narrow 3 mm diameter cylinders around the anode wires (Fig. 7.5.1.2).

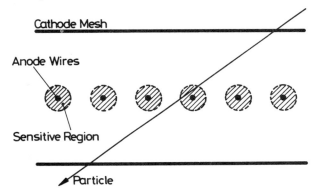

Fig. 7.5.1.2 Reduction of the sensitive region by adding electronegative vapour. For oblique track, only one wire pulses, instead of five. With a wire spacing of 0·5 cm, 40 μm wires and an anode-cathode distance of 0·7 mm, the loci of 50% efficiency had radii 1·6 mm and 2·1 mm for 59% and 21% ethyl-bromide in pentane. (Grunberg et al., 1970)

A useful notion for describing the behaviour of a Charpak chamber is the efficiency plateau. This is the region of applied voltage for which the counting rate of pulses above a predetermined threshold is constant (for, say, a beam of minimum ionizing particles). Figure 7.5.1.3 shows that it may be flat and long (>1000 volts).

7.5.2 Time resolution and dead time

The rise-time of a pulse in a Charpak chamber will be determined by the time it takes for the electrons from ion pairs to drift to the anode and form their avalanches, and by the time for the positive avalanche ions to drift from the anode. If, as in the case of X-rays, they are concentrated because the conversion electron will only have a range of a millimetre or so, they will all arrive within a few nanoseconds of each other. On the other hand, for a traversing particle, the electrons will be distributed throughout the depth of the chambers, and the time for collecting all the

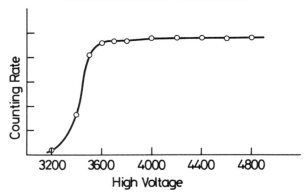

Fig. 7.5.1.3 High-voltage plateau in a multiwire proportional counter filled with argon-isobutane (80/20). $s = 3$ mm. The ordinate is the number of pulses from a wire that exceed a predetermined threshold – when a beam of beta rays is collimated on the wire. (Charpak et al., 1970)

electrons (in 7 mm) will be 150–200 nsec. Each collected electron gives its own avalanche and the final output will be the sum.

The Wilkinson expression for the rise of the output pulse ($\propto \ln(t + t_0)/t_0$) applies to the multiwire chamber because the equipotentials are circular. For 20 μm wires, 2 mm spacing, 7 mm gap and an applied voltage of 3000 V, t_0 is 1·8 nsec. Charpak et al. (1970) have calculated that although 150 μsec is required for all the positive ions to reach the cathode, 8% of the full pulse is obtained in 10 nsec.

In chambers that have a wire spacing of 2 mm, a traversing particle must travel within 1 mm of an anode wire. The maximum delay between the passage of the particle and the beginning of the output pulse is then the time for electrons to travel 1 mm in the field. It is approximately 25 nsec. The output pulse will therefore be uncertain by this amount (jitter). Experimentally, this has been checked by matching the wire pulse with that from a plastic scintillation counter (Fig. 7.5.2.1). Charpak et al. (1970) found maximum jitter times of 32, 24, 18 nsec for spacings of 3, 2, 1 mm respectively. They conclude that a time resolution of 30 nsec is obtained for 2 mm spacing, with 100% efficiency for minimum ionizing particles. The time resolution, of course, is the parameter that indicates the degree of precision obtainable in timing measurements.

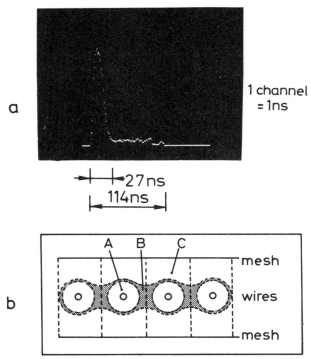

Fig. 7.5.2.1 Time jitter spectrum of pulses from one wire. The delay between the passage of a particle and the pulse appearance on a wire depends on the position at which the initiating ion pairs were produced. Those formed in region A are rapidly collected and appear in the peak, those in drift region C lie in the tail ($s = 2$ mm, $L = 4$ mm, 3·2 kV, argon-isobutane). (Steffen and Vanucci, 1969)

If an electronegative gas be introduced, the delayed pulses can be eliminated. At a cost of loss of overall efficiency the resolution can be improved by adding freon-12 and raising the discrimination threshold. Effectively, this reduces the sensitive region to a small distance around the wires. Charpak (1970) reports that with a mixture argon/freon-12 (75:25) one can obtain a resolving time of as little as 2·5 nsec at 0·7% efficiency. The limit on the practicable time resolution is really set by the specific ionization of the particles. With a wire spacing of 1 mm, the mean distance between primary ions is comparable with the 0·5 mm half-wire-spacing.

The time response may be sharpened by using several chambers in succession. For example, two chambers, each with a wire spacing of 2 mm, may be staggered so that the wires of one stand along the normal from the midpoints of the other. If the first pulse from either wire is used for timing, the jitter is much reduced. Charpak *et al.* (1970) found that FWHM jitters for one, two and three chambers were 20, 12, 8 nsec respectively. Even better resolution would be obtained with more chambers.

For an interval after a pulse has occurred, an anode wire cannot distinguish new particles. This is the dead-time for a single wire. It is governed by the time to collect all the primary electrons; by the time for the avalanche ions to drift away; and for the electronic amplifier to recover. The collection time for a 7 mm gap is likely to be ~ 200 nsec. Output pulses are typically differentiated, with times of about 50 nsec, and so the amplifier will require about 150 nsec to recover. The slow drift of positive ions is relatively unimportant. Charpak *et al.* (1970) conclude that the total dead-time need not exceed a few hundred nanoseconds. It must be noted that when maximum event rates are being considered for large chambers it is the read-out time that imposes the main limitation.

7.5.3 Spatial accuracy

The spacing (S) of the wires mainly determines the accuracy of location of a particle track. When a beam passes at right angles to the plane of the wires the uncertainty is essentially $\pm S/2$. The variance of a square distribution of width S is $S/3$. This was checked experimentally by Charpak *et al.* (1970) by passing a beam through four chambers, fitting a line with the pulses from three and noting the accuracy of the fourth. For handling and other reasons a 2 mm wire spacing is most popular, although 1 mm is certainly practicable.

Difficulties are caused when delta rays are produced in the chambers. In this event, a cluster of wires is likely to give pulses. The energy distribution of the delta rays will decide the numbers of wires that fire. The number can be as large as 10. One study found that 2% of the events gave pulses on more than four wires. It must be remembered that these clusters are asymmetrical, i.e. in general, they lie to one side of the track, and are therefore not good indicators of the track position. One can, of course, choose to reject certain clusters from one's analysis.

When the particles approach at an angle to the wire plane the depth of the chamber becomes important (see Fig. 7.5.1.2). The primary electrons will move along the field lines to the appropriate anode, with the result that many anodes will fire. For example, in a chamber with an anode–cathode gap of 8 mm (total depth 16 mm) one might expect six or seven wires to fire for a particle incident at 40° to the normal. This number can be reduced by introducing an electronic timing gate to ensure that only electrons liberated near the wires have a chance of being recorded. Also, the addition of electronegative gas is effective in reducing the collection region to a small distance from the wires [e.g. argon + isobutane + freon-13 Bl (0·3%) reduces the sensitive region to 2 mm, so that a 40° track produces pulses on only two wires—if $S = 2$ mm]. Alternatively, the spatial resolution may be improved by accepting just the signal from the first wire to pulse, on the basis that this wire will be closest to the trajectory.

7.5.4 Response in a magnetic field

In a combination of electric and magnetic fields, electrons will drift in the direction given by $\vec{E} \times \vec{B}$. Inside a Charpak chamber this drift may affect the collection of ionization electrons by a particular anode wire. The electrons may stray towards other wires. The effect is most noticeable when the wires are parallel to the magnetic field for the drift will then be in the lateral direction. No effects are seen when the electric and magnetic fields are parallel.

Charpak et al. (1970) aimed ^{55}Fe X-rays at a particular wire (of a 2 mm spacing, 8 mm gap chamber) and observed what happened when a magnetic field was switched on. At 18 kG, it was noted that the counting rate on the central wire had diminished considerably, but each adjacent wire showed enhanced counting rates. This was to be expected because the drift in the two sides of the chamber would be in opposite directions. For fields above 3 kG the drift was found to be given by the empirical relation

$$\omega = 0·9B \times 10^5 \text{ cm sec}^{-1}$$

where B is measured in kG. Fields of up to 45 kG were tested and the performance was merely blurred. The efficiency of the chamber was unaffected.

The effect may be diminished by reducing the anode–cathode gap to, say, 4 mm. In any case, for normally incident particles, the first wire to pulse should give an accurate location for the track. The investigators concluded that the effective displacements in a 20 kG field would be about 0·25 mm, which is not too serious for most purposes.

7.5.5 Performance at low pressures

Binon et al. (1971) have reported on the characteristics of proportional chambers at low pressures. Of various gases that were tested only gaseous hydrocarbons were satisfactory—in particular, isobutane, pentane and heptane. Tests were carried out in a standard CERN multiwire chamber. The pressures were varied between 3 mm and 30 mm Hg, and the voltage between 750 V and 1800 V; the value being kept 50 V below the limit of the proportional zone. The detection efficiency for 5·5 MeV alpha particles was always about 100%. Typically, across a 1 kΩ load, 5 mV pulses in isobutane and 40 mV pulses in heptane were obtained at a pressure of 5 mm corresponding to energy losses of 26 and 43 keV respectively.

The remarkable, and potentially useful, property was the time resolution. When pulses from the chamber filled with heptane (3 mm Hg) and a plastic scintillator were compared with a time-to-pulse height analyser, the excellent resolution of 2·5 ns was achieved (Fig. 7.5.5.1). The best value for isobutane was 3·3 ns and for both gases the resolution deteriorated as the pressure increased until a plateau value of about 20 ns was attained.

7.5.6 Operation in the Geiger–Müller mode

Large pulses may be obtained by operating multiwire chambers at voltages high enough to promote streamer propagation along the wires. Usefulness as a particle detector demands that the discharge must not pass from one wire to a neighbour, i.e. the mean free path of the ultra-violet light responsible for the

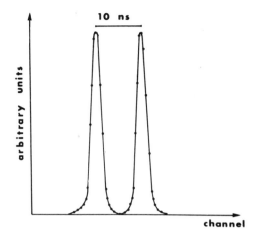

Fig. 7.5.5.1 Time distribution of pulses in a low-pressure heptane chamber (3 mm Hg, 750 V). The 10 ns shift is produced with an artificial delay. (Binon *et al.*, 1971)

streamer propagation must be very short. Charpak and Sauli (1971) have found that a suitable gas is 4% C_2H_5Br in pure argon, in which output pulses of 1 volt can be obtained on an impedance of 1000 ohm.

The actual pulse is independent of the amount of initial ionization, but the shape is a function of the site of the original avalanche; the discharge fronts advance in both directions along the wire and a discontinuity in the pulse occurs when either end is reached. The dead-time of a wire, which is the time to sweep away the positive ions, is large and likely to be about 200 μs.

The finite velocity—about 200 ns/cm—of the discharge front along the wire, opens up the prospect of determining one track coordinate by measuring the time of arrival of the discharge at a fixed point. This may be achieved with a pick-up electrode placed near, but across the wire anodes, and given a suitable potential. Charpak and Sauli found that an accuracy of 4–10 mm (FWHM) was attainable in a 38 × 38 cm chamber. And indeed a whole variety of pick-up electrode configurations may be contemplated.

7.5.7 Chamber cascades, ionization and particle identification

The fact that the energy resolution obtained with proportional counters is governed largely by the fluctuations in (a) the amplification and (b) the number of initial ion pairs, was discussed in section 7.2.3. Consequently, in multiwire proportional chambers, only a poor resolution is to be expected—as the broad ^{55}Fe X-ray and single-escape peaks indicate in Fig. 7.4.1.2. In this case a best figure of 15% was quoted.

The possibility of discriminating particles with a Charpak chamber is illustrated in Fig. 7.5.7.1 where the pulse height spectra of 370 MeV/c pions and protons are shown. The greater mean energy loss by the protons (15·3 keV vs. 3·6 keV) in the chamber gas results in much larger pulses which can easily be distinguished from the pions. The fluctuations in the electron multiplication process are large, and it is necessary to correct the experimental distribution for the finite resolution. The procedure used by Charpak et al. was to apply the measured resolution of the 5·9 keV X-ray line to the distribution after modification with the factor $E^{-\frac{1}{2}}$. In this example the particle beam was normal to the chamber, and thus a constant sensitive path was encountered by the particles. For varying approach angles, the path lengths would differ; hence, too, the number of ion pairs, and the spectra would broaden accordingly.

Identifying fast particles by measuring the energy they lose (dE/dx) in proportional counters is complicated by the large inherent fluctuations in this quantity. The fluctuations were discussed in section 1.2.3 where the relevance of the Poisson, Landau, and Blunck and Liesegang distributions was indicated. The trouble is that identical particles produce a very broad distribution of energy losses ($\equiv$ ion pairs). This means that the distributions from differing particles often overlap and are therefore difficult to resolve.

It was explained that for certain values of a parameter b^2, the Landau distribution is valid. A main characteristic of this distribution is the extended high-energy tail caused by the occasional energetic delta rays. In practice, the situation may be ameliorated by employing a stack of proportional counters. Suppose the specimen particle is made to pass through, say, N proportional counters in a row, and the pulse heights from

CASCADES, IONIZATION, PARTICLE IDENTIFICATION 285

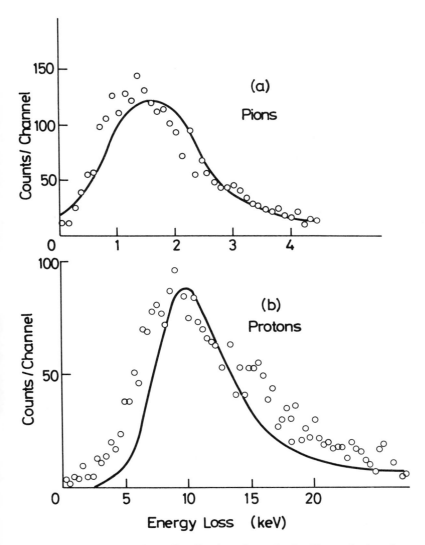

Fig. 7.5.7.1 Energy loss distributions in a single Charpak chamber by 370 MeV/c pions and protons. The points are experimental and the line is the theoretical curve corrected for finite resolution: 30% at 6 keV, varying as $E^{-\frac{1}{2}}$. (a) Pions: mean energy loss of 3·6 keV. (b) Protons: mean energy loss of 15·3 keV. (Charpak et al., 1968)

each measured. A narrowed distribution would be obtained if one then plotted the arithmetic mean of the N values. Alternatively, one might compute the average of the logarithm of the pulses (which is attractive because one can quickly obtain the logarithm of a pulse electronically). Another procedure would be to accept only the lowest of the N measured pulse heights. Yet another technique is to determine the geometric mean: the pulses from each chamber are passed through separate logarithmic amplifiers; the outputs added together; the sum fed into an antilogarithmic amplifier; and the resulting output is the geometric mean (Yuan et al., 1972).

Figure 7.5.7.2 illustrates the improvement that may be obtained by using a stack of twelve proportional counters instead of one (Ramana Murthy and Demeester, 1967). The lower diagram shows the overlap that would be measured with one counter if 100 GeV protons and pions were to obey the Landau distribution (actually, they do not). The top diagram shows how the distributions have narrowed as a result of taking, for each particle, just the lowest pulse height from the twelve counters. The curves are well separated and the particles would be identifiable with a high degree of confidence.

The "cut-off" method is an extension of this idea; a narrow distribution is obtained if the $(1 - x)N$ highest values are removed from N ionization measurements. In this case the error should be $\delta \bar{E} = \sigma/\sqrt{(xN)}$ where σ is the half width of the Gaussian-like distribution obtained by removing the Landau tail. The parameter x is chosen such that the centre of gravity of the retained measurements coincides with the position of the most probable value. This corresponds to rejecting 35% of the largest ionization values.

A third and, perhaps, best way to use a stack of counters is called the likelihood ratio method. Suppose one has a beam containing a mixture of two particles (say, protons and pions),

Fig. 7.5.7.2 Showing how resolving power may be improved by using a stack of proportional counters. (a) Landau distributions, for 100 GeV protons and pions, to be expected in a single counter. (b) The narrower distributions obtained when only the lowest pulse from twelve counters is considered. The curves are better separated than in (a). (Ramana Murthy and Demeester, 1967)

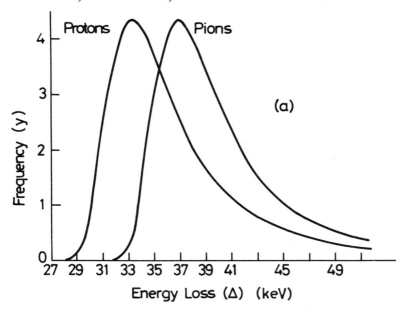

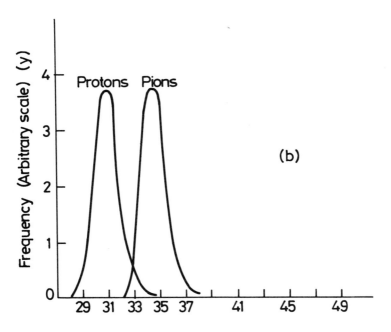

each with their own particular energy, and one wishes to distinguish them. One starts by plotting out the expected distribution curves [$y = \rho(\Delta)$, Landau, or Blunck and Liesegang] for the two particles. If, say, the beam traverses a stack of thirty counters, one computes a ratio of products for each event

$$L = \frac{\prod_{i=1}^{i=30} y_{i,\,\text{proton}}}{\prod_{i=1}^{i=30} y_{i,\,\text{pion}}}.$$

The numerator is the product of thirty terms, each being a (theoretical) ordinate of the proton distribution ($y_{i,\,\text{proton}}$) that corresponds to the observed pulse height Δ_i from counter i. The denominator is the similar product where the ordinates of the pion distribution are taken instead. If the ratio L is $\gg 1$ the particle is a proton and if $L \ll 1$, a pion. Values of $L \approx 1$ leave an ambiguity in the resolution.

These methods, developed for conventional proportional counters, will apply equally well with stacks of Charpak chambers. The success of the methods will depend on the particles under investigation. It will depend on which distribution laws are obeyed, and how well separated the individual distributions are. In section 1.2.3 we saw that relativistic particles are often described by the Blunck and Liesegang distribution which, unfortunately, is much broader than the Landau. The 100 GeV pions and protons of the above example actually obey Blunck and Liesegang, and so the optimistic statement about identifying particles (with a mean energy loss difference of 10%) with a high degree of confidence is not, in practice, justified. Ramana Murthey and Demeester showed that whereas, under Landau, 93% of protons would have lain below the minimum pion energy loss, with the real Blunck and Liesegang particles one could identify only 60% of the protons as protons, and even then there would be a 3% contamination of pions.

In spite of the obvious difficulties, the advance to higher synchrotron energies has given new impetus to this method of particle identification. At these energies ($\gamma \gg 1$) Cerenkov and time-of-flight techniques are often inapplicable and the exploitation of the relativistic rise in ionization losses becomes more profitable (Fig. 7.5.7.3). For this purpose proportional counters

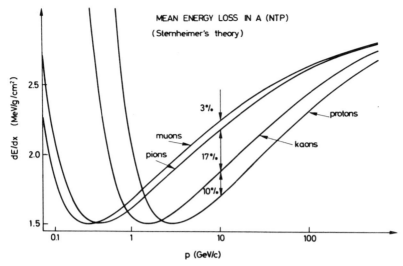

Fig. 7.5.7.3 The relativistic rise of energy losses for different particles in argon calculated from Sternheimer (1959). (Dimčovski, 1972)

have an advantage over cloud and streamer chambers, in which drops or streamers have to be counted, because the pulse height may be digitized electronically.

Following the trail blazed by Alikhanov et al. in 1956, Dimčovski (1970) and Rubbia (1970) have shown that stacks of Charpak chambers should be useful over the rise up to the plateau imposed by the density effect, i.e. over the range 5–300 GeV. It is to be noted that gas counters can best exploit the rise; the rise is very limited in solid materials with their higher densities because the polarization of atoms close to the particle's path diminishes the field acting on electrons at larger distances. Indeed Sternheimer's (1952) calculations emphasize the importance even of gas pressure; it appears that atmospheric pressure will be adequate up to perhaps 300 GeV, but reduced pressures might be necessary if the rise is to be extended, i.e. the onset of the Fermi plateau shifted, to higher energies (Fig. 7.5.7.4).

Practical considerations include the gas to be used, its pressure, the total path length and the number of layers. Some years ago, West (1953) investigated the energy loss distributions for minimum ionizing electrons in a proportional counter with various gases present. The experimental widths of the distributions, plotted as a function of probable energy loss in Fig. 7.5.7.5, are considerably larger than the Landau predictions. The widths decrease as the probable energy loss is increased, but above 10 keV little advantage is gained. The widths for different gases may be made to lie on a universal curve if

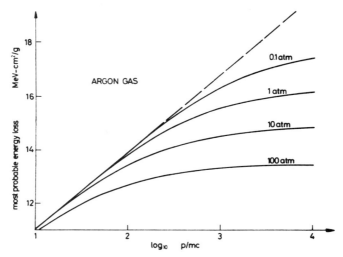

Fig. 7.5.7.4 The influence of the density effect; most probable energy losses in argon are plotted for various pressures. (Rubbia, 1970)

expressed as a function of the parameter $\xi/\overline{I_0 Z}$. The term $\overline{I_0 Z}$ is the mean ionization potential of the atoms of atomic number Z, and ξ is an energy such that, on the average, one delta of energy greater than ξ is produced in the length of track examined. Delta rays of energy approximately equal to ξ are mostly responsible for the spread in the distribution (Cranshaw, 1952). The value of ξ is given by

$$\xi = [1 \cdot 54 \ 10^5/(v/c)^2][\sum \mu_i Z_i/A_i] \text{ eV}$$

where μ_i is the mass per cm³ of the element of atomic number Z_i and atomic weight A_i.

To obtain the most compact counter, a gas should be chosen

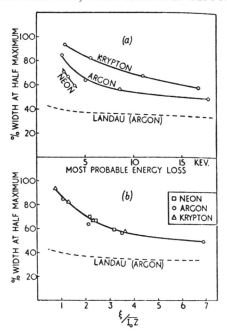

Fig. 7.5.7.5 Widths at half maximum of the energy loss distribution for minimum ionizing particles as a function of the most probable energy loss. (West, 1953)

to maximize the value of $\xi/\overline{I_0Z}$ per unit length. Values of this parameter, calculated by Rubbia (1970) are displayed in Table 7.5.7. For a given resolution, it is seen that a propane counter only needs to be one-fifth the length of an argon counter with the same number of planes. And isobutane is better still.

Using the results of a single 1 cm chamber, and applying

TABLE 7.5.7

Gas	Value of $\xi/\overline{I_0Z}$ for 1 cm at atmospheric pressure and 25°C
Argon, Krypton	0·50
Methane (CH_4)	0·970
Propane (C_3H_8)	2·58
Isobutane (C_4H_{10})	3·38

Monte Carlo calculations to derive the performance of a multi-layer detector, Rubbia proceeded to show that with a total path of one metre of propane ($\xi/\overline{I_0 Z} = 258$), no advantage is to be gained by using more than fifty layers; and that by employing the "cut-off" technique to fifty layers, relativistic pions, protons, electrons and kaons of a certain momentum could be separated (Fig. 7.5.7.6).

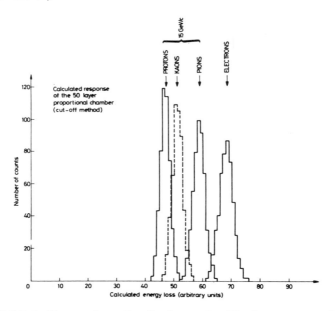

Fig. 7.5.7.6 Energy loss distributions of a fifty-layer proportional chamber system after application of the "cut-off" method. 15 GeV/c relativistic particles are seen to be separated. (Rubbia, 1970)

Dimčovski (1972) has studied the prospects for using these detectors in experiments at 300 GeV synchrotrons. Theoretically, a good separation of particles whose dE/dx differ by 10% was obtained by applying the likelihood method to 200 (and 400) layers of 1 cm. The width of the distribution increases as one travels up the relativistic rise, and Dimčovski even hinted that a separation of particles having the same, most-probable-energy-loss should be possible because the likelihood method is sensitive to the shape of the distribution curves.

Finally, Charpak chambers stacked to exploit the relativistic rise have the advantage that they can handle particles at any incident angle—unlike, say, DISC Cerenkov counters which demand a parallel trajectory.

7.6 ELECTRONIC AMPLIFIERS AND READ-OUT

It is customary to maintain the central wire-plane anode at earth potential and to apply a steady negative high-voltage (0–8000 V) to the outer cathode envelope. To prevent large currents being drawn the cathode may be connected to the power supply through a high resistance, say, 5 MΩ. At high voltages, discharges from the cathode to the outer anode wires are likely. These have been reduced by using larger diameter wires in the extreme position, although Dieperink *et al.* (1969) have preferred to line the inside of the chamber with 4 mm earthed guard strips just above and below the wires.

When long wires are used they are sensitive to high-frequency radiation. Shielding can be achieved by extending the cathode mesh to surround the wire plane in the fashion of a Faraday cage. This may also be done by coupling the mesh capacitatively to some external conducting envelope.

In order to exploit the good time resolution of proportional chambers, the electronics should be sensitive to just a few avalanches. It is therefore important to minimize the wire capacitance by reducing the conductor areas and eliminating unnecessary dielectrics.

7.6.1 Pulse amplification and preparation

The earliest circuits designed to amplify the pulses from proportional chambers were both simple and cheap. Figure 7.6.1.1 shows the three-stage amplifier by Amato and Petrucci (1968). It has a voltage gain of 200 with an output rise-time of 200 nsec. The input impedance is about 3 kΩ, and is in parallel with the capacitance of the amplifier (~ 25 pF). The output impedance is 50 Ω, to match the cable. The output pulse has a maximum of 400 mV and is negative for a negative input. With the discrimination set at -100 mV at the output, the minimum detectable wire pulse was 0·5 mV, which is quite adequate to

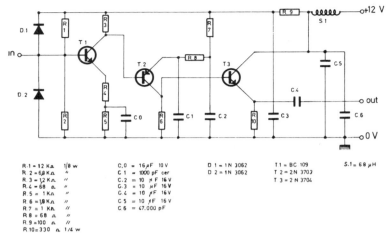

Fig. 7.6.1.1 A straightforward amplifier for proportional chamber pulses. (Amato and Petrucci, 1968)

ensure high efficiency. The pulse height spectrum shown in Fig. 7.4.1.2 was taken with this amplifier.

As an example of a more sophisticated circuit, we shall describe the design by Tarlé and Verweij (1970). It has found favour with several research groups. Commonly, the problem is to amplify a current pulse of about 500 nA (~ 1 mV across 2 kΩ, rise-time ~ 20 nsec and fall-time ~ 200 nsec) to a level of about one volt. Two cases arise. If the proportional information is to be used, the amplifier must be linear. But for many applications (e.g. hodoscopes) this need not be so.

An integrated circuit (Motorola MC–1035) containing three differential amplifiers with emitter-follower outputs and a bias driver, has been found to be a suitable amplifier (see also Koester, 1970). As explained by the manufacturer, the circuit can be used as a linear wide-band amplifier with a voltage gain of ~ 800, a rise-time of ~ 8 nsec and a maximum output of 800 mV.

Alternatively, it can be used as a trigger circuit. A sensitive monostable, with a reproducible threshold may be realized by applying positive feedback from the output to the input of the amplifier, and by controlling the amount of positive feedback with a forward-biased diode (Fig. 7.6.1.2). In the steady state

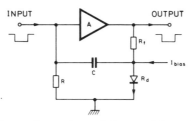

Fig. 7.6.1.2 Basic configuration of the sensitive trigger circuit. The network will be stable if the loop gain $A \times [R_d/(R_f + R_d)] < 1$. $R_d \approx 30/I_{bias}$. (Tarlé and Verweij, 1970)

the diode is biased with a current such that the feedback loop gain is less than 1. The output signal will reduce the standing current in the diode and raise its differential impedance and thus the loop gain. As soon as the loop gain exceeds 1, the circuit will switch and then behave as a normal monostable. The trigger threshold can be varied by changing the standing current in the diode.

This scheme may be performed by one differential amplifier of the integrated circuit. Although the minimum threshold was just above the noise (~ 75 μV), the minimum reproducible threshold was 3 mV. The other amplifier sections of the integrated circuit were employed as a preamplifier to the trigger (of gain 72 $\times$). Theoretically, the best input sensitivity was then ~ 45 μV, but in practice it was ~ 100 μV. The trigger worked with an output pulse width down to 75 nsec.

The complete circuit is shown in Fig. 7.6.1.3. After the trigger (third amplifier) the pulses are differentiated and inverted with

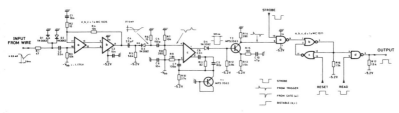

Fig. 7.6.1.3 A complete system for processing anode-wire pulses, including amplifier, trigger, gates and memory. (Tarlé and Verweij, 1970)

T_2, and are gated by the strobe signal from scintillation counters before entering the read-out logic.

7.6.2 Extraction of data

The manner in which the information is extracted from Charpak chambers depends very much on the experiment. If one is not concerned with exploiting the excellent time resolution, one may dispense with amplifiers attached to the individual wires and rely instead on a combination system. In fast decision-making applications, on the other hand, amplifiers must be employed to bring the wire pulses up to a voltage level suitable for use in fast-logic circuitry. Sometimes, too, one will wish to retain the information on the actual ionization loss.

A number of arrangements for reading out the data without the use of individual amplifiers have been described by Charpak et al. (1968). One system (Fig. 7.6.2.1) is a delay-line method in which the successive wires are connected with self-inductances. In the diagram, the time interval between the arrival of the pulses at the two ends A and B indicates the wire that pulsed. The intervals may be converted to pulse-amplitudes and recorded with a multichannel analyser. If the inductances are 2 μH each, the avalanche pulses from adjacent proportional

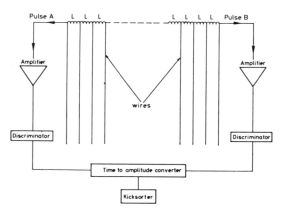

Fig. 7.6.2.1 A read-out system in which the plane of wires is built as a lumped delay-line. Pulse arrival times at ends A and B, determine the pulsing wire. (Charpak et al., 1968)

wires are separated by about 10 nsec. Rindi *et al.* (1970) have reported a similar approach using commercial delay-line for the inductances.

Another method suggested was to connect the wires with resistances instead of inductances. The relative attenuation at the two ends of the resistance chain indicates the active wire. A variant is to use resistive proportional wires; the relative attenuation at the ends indicates the position of the avalanche. Another proposal was to code the wires with an address (a, b), where (a) is the number of a group and (b) the position in the group. This coding may be achieved by appropriately coupling the wires through tiny ferrite transformers, prior to amplification.

These methods, as described above, are only really suitable for handling single pulses. Multiple simultaneous pulses would require extra, and perhaps difficult, analysis.

It appears that for most purposes it is sensible to amplify the pulses separately at each wire. As we have observed, this is especially true for decision-making applications. One factor to bear in mind is the celerity with which the pulses appear in the Charpak chambers. As a consequence, it is necessary to build in some delay so that the avalanche signals may combine properly with, say, tardy signals from scintillation counters.

A delayed pulse scheme has been described by Charpak (1970). The delay is achieved either by the insertion of a delay-line or by using the descending edge of a monostable pulse (see Fig. 7.6.1.3). The delay might be 300 nsec. The pulses may then participate in the logic that will define and record desired events.

A scheme that has the merit of eliminating individual delay units has been designed by Pagès (1970). The signals from the wires enter the circuit shown in Fig. 7.6.2.2. The gates (G_1) are normally open, and a pulse on a wire will switch the bistable memory M. The pulse proceeds through gates G_2 to the OR circuit. An output from OR closes all the gates G_1, and engages in the required decision-making function. Events will be resolved in time if the second occurs at a moment at least T seconds after the first. This resolving time can be as low as 40 nsec. During the time T, the decisions are made, and this is the dead-time—imposed on all the wires! This is a demerit because typically $T \sim 0\cdot 3$ μsec. (N.B. in the previous scheme, the dead-time only need apply to the active wire.)

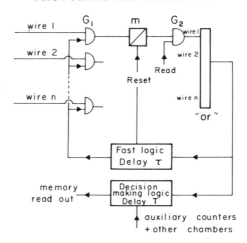

Fig. 7.6.2.2 A logic system for multiwire proportional chambers. (Pagès, 1970)

The fact is that the situation regarding the extraction of data has not yet crystallized. The most useful read-out schemes will only become known when their performance has been amply tested in experiments.

7.6.3 Read-out employing induced positive pulses

We have already noted that the negative pulse on the avalanching wire is determined by the outward drift of the positive ions. The avalanche electrons are collected instantly; the rise of the negative pulse is a result of the diminishing charge induced in the wire by the drifting ions.

A complementary positive pulse is induced on the cathode. In addition, complementary positive pulses are observed on the adjacent anode wires due to the negative-going front on the active wire. The strength of these adjacent wire pulses is, typically, only 10–20% of the active wire pulse. This means that when several wires fire (e.g. for an inclined track) the negative pulse on a wire cannot be cancelled by the positive pulse induced in the same wire. The fact that a positive pulse is obtained on both adjacent wires, even when an ionizing event occurred to one side, was interpreted by Charpak *et al.* (1970)

to mean that an avalanche always spreads completely around the wire—but not up the wire!

The positive pulses obtained at the plane cathode are a perfect reflection of the pulses at the anode. When several wires fire the cathode pulse is proportional to the energy sum of the avalanches. A pulse height spectrum for ^{55}Fe X-rays, taken from the cathode, appears indistinguishable from that in Fig. 7.4.1.2.

The pulses may be exploited in assessing an avalanche coordinate. In one scheme, Charpak et al. (1970) have placed an intermediate orthogonal wire-plane grid in a position where the equipotentials start to deviate significantly from planes. The

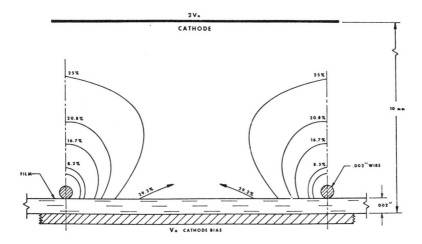

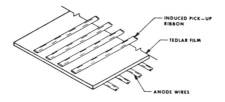

Fig. 7.6.3.1 (*Above*) An electrostatic plot of the field around the foil-supported anode wires when a continuous electrode lies immediately underneath. The wire spacing is 1 mm. (*Below*) An arrangement of orthogonal ribbons in which induced pulses will give a second coordinate of a discharge. (Neumann and Nunamaker, 1970)

location of the avalanche is then given by the negative pulse on the active wire, and the induced pulse on the orthogonal grid wire closest to the avalanche.

A potentially useful modification of the proportional chamber, that employs the positive pulses, has been devised by Neumann and Nunamaker (1970). They support 50 μm anode wires on a 50 μm thick, slightly conductive, film. They back the film with an array of fine ribbons that acts as a bias electrode. The bias voltage controls the field concentration around the anode wires. If the ribbons lie orthogonally to the anode wires, the field distribution will be approximated by the form of Fig. 7.6.3.1 (three-dimensional effects were ignored in the analogue plot with resistive paper). With the ribbons in this arrangement, the positive pulses in them may be used to give one coordinate of the avalanche.

In a test with 300 MeV/c protons the anode wires were placed 1 mm apart and the strips 0·7 mm. The authors reported that they were able to obtain spatial resolutions of 3·3 mm FWHM on the anode wires, and 1·6 mm FWHM on the orthogonal ribbons—i.e. a better resolution on the ribbons! It must be noted, however, that the efficiency was only 85% and the presence of the film spread the discharge to a dimension of about 3·5 mm.

CHAPTER 8

A Miscellany of Chambers

8.1 WIDE-GAP CHAMBERS

8.1.1 Prospects for a laser-excited chamber

The interaction of laser beams in gases has fascinated physicists in recent years. For example, the discovery by Terhune (1963), that the focused output of a Q-switched infra-red laser could cause ionization and breakdown, has led to much speculation. Grey Morgan (1967) has discussed several possible mechanisms: the liberation of a free electron in helium might involve the simultaneous absorption of 14 photons (ruby laser, 6934 Å = 1·7 eV), or perhaps the unquantized radiation theory of Davidson (1947) needs to be invoked; and the discharge might proceed by a cascade inverse bremsstrahlung process which would be essentially an extrapolation to optical frequencies of known microwave discharge theory.

If one wishes to devise a laser-excited track chamber, the initial electrons must be those liberated by the ionizing particle: the laser beam merely has to cause electron multiplication. The advantage to be gained in applying a very high-frequency (infra-red) field is that the discharge should be severely limited in volume—the space-charge effects that lead to the rapid expansion of streamers should not obtain.

In a vacuum and under the influence of an alternating field, a free electron oscillates with its velocity 90° out of phase with the field. The electron can only gain energy by colliding with gas atoms, when its oscillatory motion is changed to random motion (Brown, 1959). After many collisions the electron may have sufficient energy to make an inelastic collision, i.e. it may excite or ionize a gas atom. The rate at which the electron gains energy from the alternating electric field (frequency ω, amplitude E_0) is

$$P = \frac{e^2}{m} \frac{\nu}{\omega^2 + \nu^2} \frac{E_0^2}{2}$$

where ν is the collision frequency.

In a calculation of the requirements for a laser-activated

chamber, Schneider (1963) has given the following argument. If the average energy required for each ionization is U_i (volts), the time constant for electron multiplication is

$$\tau = U_i \frac{e}{P}.$$

The production rate of photons in a gas discharge is of the same order as the rate of ionization, and thus the time required to obtain a multiplication of e^α will be

$$t = \alpha\tau = \alpha U_i \frac{e}{P}.$$

During this time the plasma ball, produced by the initial electron will grow by free electron diffusion to a radius roughly estimated to be

$$r = \lambda\sqrt{(t\,v)}$$

where λ is the electron mean free path. (N.B. ambipolar diffusion would give a lower value.) Now the energy flux per unit area of an electromagnetic wave is

$$\dot{P}_{em} = c\varepsilon \frac{E_0^2}{2}$$

where ε is the dielectric constant, and substituting,

$$\dot{P}_{em} = \alpha U_i c\varepsilon \frac{m}{e}(\omega^2 + v^2)\left(\frac{\lambda}{r}\right)^2.$$

In practice the laser beam frequency ω will be much larger than the collision frequency v, so

$$\dot{P}_{em} = 2\alpha \frac{U_i U_e}{r^2}\sqrt{\frac{\varepsilon}{\mu}}\left(\frac{\omega}{v}\right)^2 \quad [\text{Watt m}^{-2}]$$

where U_e is the average electron energy in volts; and the total energy per unit area of the light beam must be

$$E = \alpha \frac{m}{e} U_i \sqrt{\frac{\varepsilon}{\mu}\frac{\omega^2}{v}} \quad [\text{Joule m}^{-2}].$$

Taking reasonable values ($\alpha = 25$ to obtain a visible spot from one primary electron; a plasma ball radius of 1 mm; $U_i = 50$ volts; $U_e = 10$ volts; and a collision frequency $3 \cdot 10^{14}$ Hz corresponding to 10 atm pressure) we find

$$P_{em} \approx 10^6 \quad [\text{Watt cm}^{-2}]$$

and

$$E \approx 50 \quad [\text{Joule cm}^{-2}].$$

Schneider concluded that it should be possible to sweep a volume of 100 litres with a laser of 1 M Watt output power, if one could reflect the beam up and down the chamber with a reflexion coefficient of 0·999.

Gygi and Schneider (1969) have made some preliminary studies on laser activation. They mounted a ^{90}Sr source within a high-pressure discharge chamber equipped with a scintillation counter for detecting beta particles. The chamber could be illuminated with a 30 Joule (15 mm diameter beam, 7 nsec pulse) laser which was fired by a Kerr cell on recognition of a beta particle.

No amplification was detected in henogal (neon/helium) even at 20 atm; but in argon, starting at 14 atm, tiny white spots appeared on 3000 a.s.a. film at $f/16$. At higher pressures large luminous areas were observed. The authors tentatively concluded that single electrons could result in tiny (0·1 mm) spheres which would radiate about 10^{11} photons in the visible region. When laser technique has advanced, it may be possible to realize a track chamber with a very high spatial precision.

8.1.2 Microwave discharge chamber

A number of attempts have been made to produce visible tracks in chambers energized with microwaves. The aim is to confine luminous discharges to the ionization sites along a trajectory; the high-frequency oscillating field should cause electron multiplication, and the direction of avalanche development should alternate, to and fro, according to the field.

In an early trial, Bevan (1949) coupled an 8 cm diameter, wire-gauze waveguide to a 60 kW, 3 GHz magnetron which provided 1·4 μsec pulses. Standing waves were avoided with a suitable termination. A glass tube containing neon saturated with iodine vapour (total pressure 20 cm Hg) lay within the waveguide. It was found that diffuse discharges corresponding to alpha and beta particles could be seen and photographed through the gauze.

Lederman (1961) reported applying a 9 GHz magnetron to a rectangular resonant cavity. This cavity had a volume of 7 cc, a gap of 3 mm, was filled with argon and was operated in a TE_{202} mode. The magnetron pulse was triggered by scintillation counters (delay 500 ns), and discharges could be viewed through a perspex window in the cavity. It was found that a cosmic particle passing through the chamber caused vertical sparks at the antinodes in the cavity. A dead-time of 20 msec was achieved with a sweeping field.

In another study, Fukui *et al.* (1963) applied a 5 GHz, 250 kW, 0·1 μsec pulse, C band magnetron to a 4·7 × 2·2 cm² waveguide in which lay a glass tube containing neon + 1% argon + 0·4% $HCOOC_2H_5$ to a total pressure of 500 mm. The resulting electric field was 2 kV/cm. With one side wall replaced by a wire grid, photographs of tracks could be taken (Fig. 8.1.2.1).

Fig. 8.1.2.1 Particle track in a microwave discharge chamber. (Ohasi, 1966)

No experiment appears yet to have been performed with a microwave chamber. Significant handicaps are that the size of the sensitive volume is small, being limited by the microwave frequency; and that it is difficult to achieve a uniform field. Fukui *et al.*, have suggested that a lower frequency, say 1000 MHz, and a circularly polarized wave might be capable of activating a 20 cm diameter cylinder, many tens of centimetres long. Another approach is that by Doviak *et al.* (1967) who folded the waveguide in such a way that the particle traverses it several times, and therefore gives several photographable discharges along the trajectory.

8.1.3 Chamber with a rotating electric field

One drawback of the wide-gap chamber operated in the merged streamer regime is that it is not isotropic: tracks will be reliably delineated only if they lie within 45° of the electric field [see section 6.3.1]. Koslov and Rudenko (1969) have attempted to circumvent this handicap by varying the direction of the electric field within the chamber during the formation of the streamer channel.

In their early work (1968), they straddled the four sides of a glass box with two pairs of plate electrodes. On recognition of a particle trajectory, a pulsed field was applied with one pair and this was immediately followed by a pulsed field at right angles, provided by the second pair. During the first limited pulse, if the angle be less than 45°, isolated streamers will result along the track. During the second pulse, the isolated streamers will merge to create a spark-like channel and the merged trail (originally at $<45°$, now $>45°$) will remain relatively unaffected because the high ionization will prevent electrons being extracted. Thus in the plane of the field directions, bright tracks will occur regardless of their orientation. This will hold for angles up to about 45° with respect to this plane. Kozlov and Rudenko calculated the sensitive solid angle to be 3π.

Initially, pairs of plate electrodes were used. However, it was found that their mutual proximity gave rise to a distortion of the field in the chamber. An alternative design was chosen, consisting of brass rods lying within the cavity of a double-walled perspex box and along the line of its four edges. Pulses were applied to diagonal pairs of rods and the fields were thus diagonally oriented. With the wall cavity filled with glycerine, the resistive/capacitative coupling ensures that a uniform potential distribution is maintained down the walls for the duration of a pulse, and a satisfactory field results. Each electrode pair was connected via a cable to the single output of a Marx generator. (Actually, the Russians call them Arkad'ev–Marx generators.) The pulse across one pair was delayed relative to the other by increasing the length of its connecting cable; and the pulse lengths were limited independently with shunting spark gaps across each pair of electrodes. Satisfactory particle tracks were obtained with this double-pulsing technique when

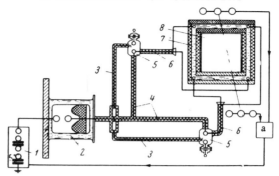

Fig. 8.1.3.1 Chamber with a rotating electric field. Suitably phased alternating voltage trains are applied to the diagonally placed pairs of rod electrodes. (Kozlov and Rudenko, 1969)

the box was filled with neon (750 Torr) and saturated ethyl alcohol vapour.

In a further development, Kozlov and Rudenko (1969) rotated the electric field vector in the chamber, round and round, by applying sinusoidally decaying trains of pulses to the pairs of electrodes. Their arrangement is shown in Fig. 8.1.3.1. The pulses were derived from the 500 kV, 2 ns rise-time Marx; sharpened with the capacitor (2) and series gap; split; and fed through cables (4, 6) to the orthogonal pairs of rod electrodes mounted along the edges of the chamber. The alternating voltages on the chamber (Fig. 8.1.3.2) were produced by limiting the original pulses with discharge gaps (5) and allowing multiple reflections to occur along cable (6) between the chamber and the gaps (5). The period of the voltage was $4l/v$, where l is the length of cable (6) and v is the pulse propagation velocity in the cable.

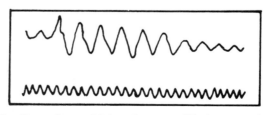

Fig. 8.1.3.2 One of two high-voltage oscillations used to create a rotating electric field. The (lower) calibration oscillation has a period of 10 ns. (Kozlov and Rudenko, 1969)

Shunt resistances across the chamber electrodes determined the rate of decay—typically, the oscillations had disappeared in 3 μsec. Each of the discharge gaps (5)—operated at 5 atm nitrogen—incorporated an ancillary gap which primed the main gap with ultra-violet. This priming, time-controlled via cable (3), ensured a very stable pulse length in the range 7 ± 0.5 to 29 ± 3 nsec.

The rotating electric field vector resulted from a suitable phase difference—adjusted by altering the length of cable (6)—between the two voltage oscillations. In the chamber, the ionization along the particle's path initiated avalanches which developed in a circular fashion. These merged to create a continuous ionized channel and handsome tracks at many orientations were photographed. Kozlov and Rudenko claimed a good performance over a solid angle of 3π.

8.2 HYBRID PROPORTIONAL/SPARK CHAMBER

A multi-electrode chamber which includes distinct spark and proportional mode functions has been devised by Fischer and Shibata (1970). The resulting instrument offers many advantages, including a reasonable cost; the timing characteristics of a proportional chamber; and the spatial resolution of a digital spark chamber.

The disposition of electrodes and voltages in the hybrid chamber is shown in Fig. 8.2.1.1. Ionization electrons from the

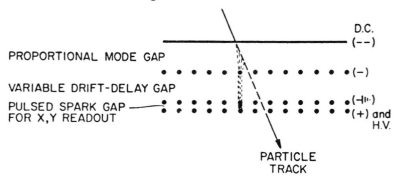

Fig. 8.2.1.1 Hybrid chamber. Avalanches form in the proportional gap; an electron cloud drifts into the narrow gap to initiate sparks on command. (Fischer and Shibata, 1970)

track of a traversing particle drift in the proportional gap and form avalanches at the anode wires. A large number of the avalanche electrons escape and drift through the delay gap; they enter the final narrow gap and normally vanish on striking the last electrode. However, if during the time of passage of electrons through the delay gap, external counters and their associated logic have recognized an interesting event, a high-voltage pulse is applied to the final gap while the electrons are drifting in it. A spark is formed and x and y coordinates can be ascertained in a conventional fashion, e.g. by magnetostriction. The voltages are chosen so that sparks only form from electron clouds, and not from the few electrons liberated along a particle track.

The time resolution of the chamber is determined by the drift time of the track electrons in the proportional gap. Unlike the spark chamber which stores all the tracks acquired in the delay before triggering, the proper time sequence of events is preserved. The recovery time is shorter than in spark chambers because the narrower gap permits the rapid clearing of ions left by a spark. The fact that the proportional wires can be closely spaced (e.g. 50 μm wires, spaced at 0·6 mm) implies a good spatial resolution: this close spacing is facilitated by using gases with a fast electron drift velocity—Fischer and Shibata recommend spark chamber neon (Ne + 10% He) plus 70 Torr methyl alcohol.

In tests, with gap sizes of 6·3 and 1·6 mm, resolving times of about 75 nsec (FWHM) were obtained, with a recovery time of 80 μsec and a spatial resolution of 0·3 mm. A similar performance has been obtained with hybrid chambers developed at Karlsruhe (Bohmer and Schopper, 1971) and repetition rates of up to 10^4 per second are predicted for experimental operation.

8.3 NEON FLASH TUBES

An assembly of neon flash tubes constitutes a large volume detector that is simple to manufacture and to operate. They have played an important role in cosmic-ray studies both at sea level and underground. The method, which was first demonstrated by Conversi and Gozzini (1955), involves the mounting of a large number of neon-filled glass tubes between plane electrodes.

When an ionizing particle is recognized to have traversed the assembly, the application of a high-voltage pulse to the electrodes causes just the tubes along the trajectory to flash. Photographing the array end-on enables a record of the track to be taken (Fig. 8.3.0.1).

Cosmic-ray experiments are slow and any apparatus must have a long and stable life. Care must therefore be taken in the construction and filling of flash tubes. Hampson and Rastin (1971) made tubes from 0·5 cm diameter (internal) soda glass. After evacuation down to 10^{-5} mm Hg, followed by flaming to remove absorbed gases, the tubes were filled with commercial neon. Tests show that better efficiencies are obtained with higher neon pressures—3 atm seems to be a good choice. Each tube should be painted white to improve the axial light emission, and then given an outside layer of black paint.

Coxell and Wolfendale (1960) investigated the characteristics of an assembly of flash tubes. The efficiency of a tube declined when the delay prior to applying the high-voltage pulse was increased: $\sim 100\%$ for 1 μsec; $\sim 90\%$ for 10 μsec; and $\sim 1\%$ for 100 μsec. This decline must be explained by the removal of free electrons from the trail; likely causes are: recombination; diffusion to the walls; negative impurity ion formation; and sweeping by residual fields of remnant charges left by previous discharges. The flash tubes exhibited a dead-time; if sufficient

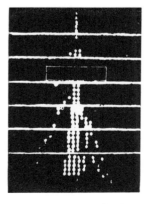

Fig. 8.3.0.1 Particle tracks manifested by neon flash tubes. (Conversi and Gozzini, 1955)

time was not allowed to elapse before the application of a second high-voltage pulse, spurious "after-flashing" occurred. The probability for after-flashing was about 80% for a 20 msec delay, and approached zero at 100 msec. In a test of the lifetime, a single tube was flashed 10^6 times, during which its efficiency declined from 97·5 to 94·5%. The spatial resolution of one tube is limited by its diameter; but if many layers of tubes are used, the precision of location can be enhanced. Coxell and Wolfendale mention an uncertainty of 0·46 cm when five layers are used; more layers would be even better and, of course, would also ensure 100% efficiency for recording particles.

A feature which should not be overlooked is that a large volume of flash tubes responds to particles incident at all angles. Wolfendale's Collaboration (Menon et al., 1967) in the Kolar Gold Field cosmic neutrino experiment successfully exploited such large detectors. Photographic recording was used. Reines (1967) reported the installation of 60,000 Conversi tubes (each 1·8 cm diameter, 2 m long) in his neutrino experiment. At the end of each tube was situated a light sensor which was connected to a complementary bulb on a display board—to be recorded on video tape. As an alternative means of recording, Bacon and Nash (1965) proposed mounting two wire electrodes down the inside of each tube. These wires maintained a steady orthogonal field of 3 kV/cm, and in the event of the tube flashing, a spark occurs whose current can be detected externally.

The discharge mechanism in a neon flash tube has been investigated by Hampson and Rastin (1971). It appears the processes differed, according to the pressure. When the filling was one atmosphere, the few primary electrons initiated avalanches that readily developed into streamers. On the other hand, at the higher pressures (3 atm), the primary avalanches did not develop into streamers; rather, they caused other avalanches in the tube—presumably by photoionization. In this case, the discharge spread to fill the whole tube. Only after this had happened—i.e. when a high density of free electrons had been built up—did streamers begin to appear. Thus at the higher pressures, streamers are a consequence of the discharge and not its cause.

Conversi (1973) has proposed using flexible plastic tubes as flash tubes. A complete chamber might consist of many such tubes bent in a configuration to suit the experiment. Thus in a design for an experiment on Adone, the electron-positron storage ring at Frascati, a chamber is proposed to enclose the vacuum pipe. A single cylindrical layer of large area is constituted by dozens of contiguous semicircular flash tubes.

8.4 HYBRID-EXPANSION CHAMBERS

8.4.1 Streamer-cloud chamber

A successful combination of streamer and cloud chamber technique, in which photographable droplets condense on avalanches initiated by ionization electrons, has been demonstrated by Manjavidze and Roinishvili (1967). The benefits are that a time resolution of a few microseconds is combined with a long memory time for selected events.

The apparatus is shown in Fig. 8.4.1.1. When a particle's passage is recorded by counters (7), the Arkadiev–Marx

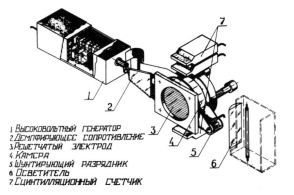

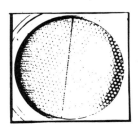

Fig. 8.4.1.1 A cosmic-ray track taken with the Tblissi streamer-cloud chamber. (*Courtesy of V. N. Roinishvili*)

generator (1) transmits a high-voltage pulse to the chamber which is filled with neon (1·1 atm) and ethyl alcohol. The pulsed field, limited with a shunting spark gap, produces avalanches along the trajectory. With such large concentrations of ions acting as condensation nuclei, a chamber expansion of only 1·04 was required. After a delay of 40 msec, a 50 Joule pulsed light source illuminated the drops. A photograph of a cosmic-ray track is shown in the figure; it was taken along the direction of the electric field in the chamber.

8.4.2 Towards the triggered bubble chamber

Every undergraduate knows that the bubble chamber has the grave disadvantage that it cannot be triggered by external electronic counters set to register a certain combination of emerging particles. Thus, in normal operation, the expansion cycle of the chamber is synchronized with the burst of particles from an accelerator; in a series of exposures, desired events may occur as a matter of chance—uninteresting photographs are rejected.

The author (1973) has speculated that since twenty years have elapsed since the invention of the bubble chamber it is timely to reconsider the matter of triggering in the light of recent advances in technology. The incentive to do so lies in the new promise of high-energy experiments on cosmic rays, and at intersecting storage rings, where random activation of bubble chambers is likely to prove useless.

Although the theory of bubble formation is incomplete, the Seitz (1958) model has found acceptance as a qualitative picture (Peyrou, 1967). It assumes that bubbles are formed by thermal spikes produced by the heat released in ion recombination at the end of delta rays. Enough thermal energy must be released within a volume corresponding to the critical radius for bubble growth—typically 80 Å—to provide the necessary surface energy, evaporation energy and work against the pressure. Quantitatively, discrepancies with experiment occur; the observed bubble density (c. 15/cm) is much lower than the density of apparently capable delta rays. Experiments correlating the relativistic increase of ionization with bubble density lead to the conclusion that delta rays initiate bubbles with an efficiency that is practically independent of their energy (if above a certain small value) and which amounts to only 1 or 2%. It appears that this small value—i.e. the necessary energy for bubble formation—is 10–40 eV for hydrogen, and about 200 eV for heavier liquids.

Hydrogen, and many other liquids, have proved successful bubble-chamber media. In a case of special interest to us, liquid xenon, satisfactory operation has been achieved only with the addition of about $\frac{1}{2}$% ethylene (Brown et al., 1956; Kanarek et al., 1959). It seems that xenon is a good scintillator, and that a quenching agent is necessary to ensure the local release of the recombination energy as heat. It was partly this fact which led to the downfall of the early Glaser theory—that the energy for bubble growth derives from the electrostatic repulsion of ions—for the electrostatic mechanism would be as good in xenon as any other liquid.

In conventional operation the susceptible condition of superheat is produced by pressure reduction; an expansion mechanism is activated about 10 msec before the arrival of the beam. It is this delay that has prevented the development of a triggered instrument; the cooling of the embryonic bubbles has been calculated to occur within about 10^{-10} sec of the deposition of energy by the delta ray.

The author has conjectured that the recent work on Blumleins (see section 6.2.2) in which voltages up to 10 MV have been obtained, and on avalanche multiplication in liquid xenon (see section 8.5.3), might offer the prospect of a triggerable bubble chamber.

Xenon is a noble liquid and in it free electrons remain free, i.e. they do not become attached to form negative ions. Thus electrons liberated by an ionizing particle in a chamber filled with pure xenon will continue to exist for milliseconds along the line of the track (ignoring, for the moment, diffusion and some recombination). In considering the prospect for triggering, the question arises whether these electrons could be activated, say, 5 msec later, by which time the liquid could be brought to a superheated state. And if at this late moment, a Marx pulse were applied to electrodes in the chamber subjecting the whole volume to a field of 1 MV/cm, would bubbles grow on the resulting avalanches?

Perhaps they would. Let us contemplate the system shown in Fig. 8.4.2.1; it is really a hybrid streamer-cum-bubble chamber. A logic signal from the counters initiates a rapid expansion. At the moment of superheat, a 2 MV pulse is applied to alternate electrodes in the chamber. If bubbles grow

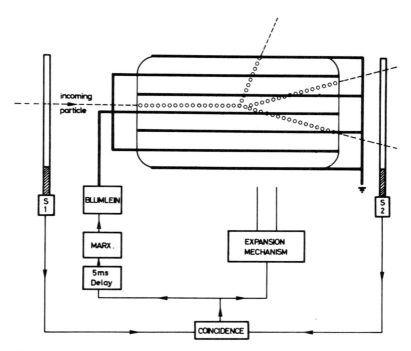

Fig. 8.4.2.1 Sketch of a proposed xenon bubble chamber and its trigger system. (Rice-Evans, 1973)

on the avalanches, they could be illuminated through the gaps and photographed in a conventional manner. The spacing would have to be 2 cm to achieve a field of 1 MV/cm. At these high fields, field emission of electrons could be awkward, and it would be necessary to coat cathode surfaces with a thin layer of insulator, e.g. mylar which can withstand fields up to 5 MV/cm. Larger gaps could be used if either higher Marx voltages were applied or if uniform fields of less than 1 MV/cm would suffice.

Whereas in the Berkeley case the avalanche must have been restricted to a few μm length adjacent to the wire, in a uniform field no such limitation obtains. In a field of 2 MV/cm, the Townsend coefficient is 4×10^4/cm (Müller *et al.*, 1971) and a gain of 4000 might be achieved in an avalanche two microns in length and developed in ~ 2 ns. Making reasonable geometrical assumptions, the electron density at the avalanche head is about 3×10^{23} m^{-3}, which corresponds to 0·6 electrons in a critical bubble of 2×10^{-24} m^3. Taking the saturated electron drift velocity as 3×10^5 cm/s (Miller *et al.*, 1968), the eighteen electrons will sweep the bubble diameter and deposit 60 eV in acoustic collisions during the heat dissipation time of 10^{-10} s. This figure approaches the 200 eV for bubble growth, and in any case would be subject to statistical fluctuations—the specific ionization in xenon is 20 ion pairs/μm (Zaklad, 1971). It would also be enhanced by energy released in recombination.

The early experience, mentioned above, where it was found that for bubble growth pure xenon needed a quenching agent thermally to localize the energy of recombination, raises some doubts. However, with a large ion multiplication, the density of recombining ions—in the heads of the avalanches—may be much larger than in the conventional xenon chamber; and the enhanced recombination alone might liberate enough thermal energy within critical radii to initiate bubbles. One cannot make proper quantitative predictions, partly because not enough is known about recombination. And, of course, the energy release calculated above is largely mechanical and hence a quencher would not be needed anyway.

Some technical development will be needed for the proposal to succeed; attention will have to be paid to the creation of, say, 20 ns, 2 MV pulses, and their application to a capacitance considerably larger than that presented by a streamer chamber. The dielectric constant of liquid xenon is unknown; we assume it to be near unity. A small continuous clearing field would have to be imposed throughout the chamber to sweep away old tracks; it could be switched off within 50 ns of a trigger.

On the question of diffusion: in the 5 ms interlude prior to sensitization the mean electron displacement away from the track ($\bar{x} = \sqrt{(2Dt)}$) would be 4 mm—a most serious handicap! The solution might lie in finding a suitable quenching compound that, when added to xenon would attach free electrons but be content to liberate them when the field is applied (Kotenko *et al.*, 1967). Double pulsing might help: a high-voltage pulse is applied within, say, 400 ns of the particle's traversal; and the second, of opposite polarity, at the moment of superheat. The avalanche due to the first would ensure the survival of the ions during the expansion, and keep their distribution centred on the track, and the later oppositely-directed avalanche/streamer might deposit the required thermal energy. (Double pulsing with two Marx has been achieved in another context—Eckardt, 1970) (see Fig. 8.4.2.2).)

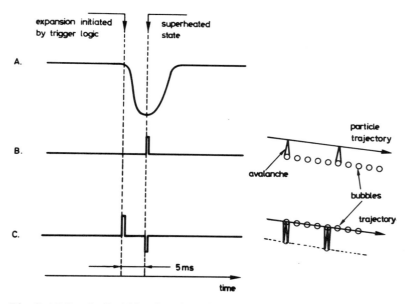

Fig. 8.4.2.2 A. Bubble-chamber decompression cycle. B. A single HV pulse applied at the moment of superheat – hypothetically giving line of displaced bubbles. C. Two HV pulses of alternate polarity – hypothetically restoring the bubbles to the trajectory. (Rice-Evans, 1973)

It may be concluded that the prospect for a triggerable streamer-cum-bubble chamber is sufficiently promising to warrant some practical study.

8.5 NOBLE LIQUID CHAMBERS
8.5.1 Prospects for liquids

Many advantages would accrue if multiwire proportional counters could be successfully operated with liquids rather than gases. The higher density should permit the use of thin gaps and a close wire spacing. This follows from the fact that whereas 110 ion pairs are produced in 1 cm of argon at S.T.P., the total specific ionization in liquid argon is 10^5 pairs/cm. Lateral uncertainties due to drifting and delta rays would be reduced and a high spatial resolution might be expected.

The success of noble gases in counters and chambers is a consequence of their not being electronegative: electrons may drift in them freely without becoming attached. In view of this, it is natural to contemplate the noble elements again as candidates for an operating liquid. Table 8.5.1 shows that they will be accompanied by cryogenic problems.

TABLE 8.5.1

Element	Melting point (°C)	Boiling point (°C)
Helium	−272	−269
Neon	−249	−246
Argon	−189	−186
Krypton	−157	−153
Xenon	−112	−107

Some early work with liquids should be mentioned. For example, in 1961 Swan investigated spark breakdown in liquid argon. In 1963 Reibel and Schluter attempted to make a high resolution optical spark chamber using liquid helium in a 0·127 mm gap between parallel plates—but they observed no track sensitivity when a 32 kV pulse was applied 200 nsec after an event. In 1968 Riegler obtained sparks in a 2 mm liquid argon gap with 100 kV pulses, 10 nsec rise-time and 400 nsec delay. To prevent recombination a 2 kV clearing field was applied. A vague correlation between tracks and sparks was found.

8.5.2 Electron multiplication in liquid argon

Alvarez and his team at Berkeley (Derenzo et al., 1969) have studied the action of cylindrical counters filled with liquid argon. Initially, a charge-sensitive amplifier with a noise level of 10^{-3} picocoulombs was connected to a 13 μm anode wire that lay within a 4 mm diameter cathode. A stopping 1 MeV electron should deposit 30,000 ion pairs (5×10^{-3} pC), and indeed, corresponding pulses were detected merely by collecting this charge. Actually, the counter was irradiated with 1 MeV gamma rays from zinc 65.

When a 4 μm tungsten wire was used with a field of 1.5×10^6 V/cm at its surface, large pulses with a uniform height were discovered. Subsequently, these pulses—having a rise-time of 20 μsec and a decay-time of 100 μsec—have been seen with both 8 and 13 μm wires. The constant pulse size (1–50 pC, depending on the applied voltage), regardless of the Compton distribution of electrons from the gamma rays, indicated that proportional amplification was not occurring, but rather, a Geiger–Müller-type mechanism was coming into play.

The appropriate drift equations are: for positive argon ions, $v_+ = k_+ E$ where $k_+ = 2.8 \times 10^{-3}$ cm^2/Vsec; and for electrons $v_- = 3160\ E^{\frac{1}{2}}$. Recombination of ion pairs is significant: the fraction evading recombination is $1/(1 + K/E)$ for minimum-ionizing particles, where $K = 1100$ V/cm (Marshall, 1954). In liquid argon, Swan's (1961) values for the Townsend coefficient α are the best available: $\alpha = 5 \times 10^9 \exp -(2.27 \times 10^4/E^{\frac{1}{2}})$.

In their consideration of the processes, Derenzo et al., pointed to the relevance of excitation collisions. Starting with one electron, the first avalanche creates, say, M electrons ($M = \exp(\int \alpha\, dr)$) and N excited argon atoms. Each excited atom has a probability p of producing a secondary electron through either (i) ultra-violet emission liberating a photoelectron from the cathode or an impurity molecule; or (ii) through an ionizing collision with an impurity molecule. The second Townsend coefficient ($\gamma = Np/M$) is the average number of secondary electrons released for each of the M electrons in the previous avalanche. $M\gamma$ electrons are thus liberated and their avalanche results in $M^2\gamma$ electrons. The process continues and the grand total is given by:

$$M + M^2\gamma + M^3\gamma^2 + \ldots = M/(1 - M\gamma) \quad \text{(for } M\gamma < 1\text{)}.$$

If $M\gamma$ is greater than 1, the series diverges and the gain is infinite. In practice the pulse size will be limited by the positive ion space-charge which reduces the field at the anode. The authors suggest that the absence of a sizeable proportional region indicates a large value of γ: rather than 10^{-3}, a value of ~ 0.5 might explain the lack of intermediate pulses. (N.B. An implication is that the mean free path for ionizing collisions is 0.2 to 0.3 μm.)

Consideration has to be given to the smoothness of the wires. The Berkeley team found that Ni–Fe and stainless-steel wires, drawn through twenty-five diamond dies, gave better results than tungsten or other stainless-steel wire. With rough wire, pulses appeared at 1.5×10^6 V/cm and sparking at 1.8×10^6 V/cm; with the smoother wires, pulses occurred at 3×10^5 V/cm [*sic*].

However, on the basis of further experience, it was conceded that the prospects for a liquid argon counter were dim because either they were too inefficient (detecting $<20\%$ of particles) or they required the heating of a thick 100 μm wire (which would preclude high spatial resolution) (Müller *et al.*, 1971).

8.5.3 Experience with liquid xenon

When the Berkeley group transferred their attention to another noble liquid, xenon, the prospects brightened (Müller *et al.*, 1971). Proportional multiplication was seen with anode fields $>10^6$ V/cm, and alpha particles were detected with 100% efficiency. Their chamber, composed of glass, kovar and stainless steel, is shown in Fig. 8.5.3.1. A 3.5 μm anode wire was mounted within a cathode composed of a conductive SnO coating on the inner, 8 mm diameter surface of the glass cylinder.

It was found that for anode fields up to 3×10^6 V/cm, the electron multiplication in liquid xenon resembled that in gas-filled counters. Figure 8.5.3.2 shows a pulse height spectrum for 5.5 MeV alpha particles corresponding to a gain of 20. As the voltage is raised, the gain increases up to about 100, after which a new phenomenon occurs: pulses appear with heights (1–4 pC) that are independent of the initial ionization (Fig. 8.5.3.3). It is interesting that whereas the proportional pulses rise in about 150 nsec (consistent with mobility $k_+ = 3 \times 10^{-4}$ cm^2/Vsec),

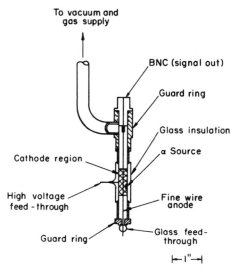

Fig. 8.5.3.1 Single-wire counter for liquid xenon studies. (Muller et al., 1971)

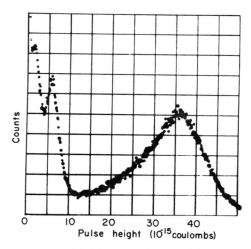

Fig. 8.5.3.2 Spectrum of 5·5 MeV alpha-particles in the liquid xenon counter at 2300 V (liquid gain ~ 20). The Am^{241} source was attached to the cathode. Recombination has reduced the initial ionization by a factor of 15 and broadened the peak. The left-hand peak corresponds to 60 keV gamma-rays. (Muller et al., 1971)

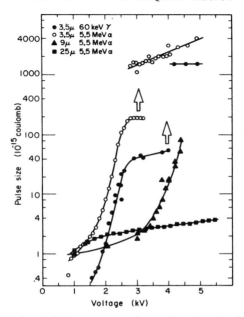

Fig. 8.5.3.3 Pulse height versus voltage for the single-wire xenon chamber for various anode-wire diameters. (Muller *et al.*, 1971)

the picocoulomb pulses rise in less than 30 nsec. The reason is not clear: anomalous ion mobilities in high fields, or ultra-violet action may be responsible.

Extraordinary care was required to remove electronegative impurities; the presence of only a few parts per million could seriously impair the performance. The xenon gas was purified by circulating for hours over a hot (200°C) copper catalyst preparation, and a cold (−78°C) molecular sieve 4A (Zaklad, 1971). The gas was liquefied when the chamber was immersed in a cold freon-11 (CCl_3F) bath. (N.B. freon-11 is liquid at 20°C and −107°C.)

The high-specific ionization in liquid xenon means that electron multiplication is not essential for the creation of measurable pulses. Müller *et al.* reported that a multiwire anode plane (12 μm wires, spacing 50 μm) situated 0·7 mm from a plane cathode, would function satisfactorily as an ionization chamber. With a charge-sensitive amplifier attached to one of the anode wires, a spatial accuracy of $\sigma_{rms} = 25$ μm, was obtained in recording alpha particles. In a later report (Derenzo *et al.*, 1971) a value of $\sigma_{rms} = 15$ μm was obtained with a wire spacing of 25 μm. It must be concluded from these early experiments that the liquid xenon counter holds great promise for locating particle trajectories with a high precision.

8.6 SPARK COUNTERS

In the early literature, articles may be found referring to spark counters. They refer to arrangements of two electrodes with a steady high voltage maintained between them. For example, Greinacher (1934) showed that strongly ionizing particles could initiate sparks which could duly be registered. The action of any particular counter depends on the geometry of the electrodes and the choice of operating gas. In one configuration (Chang and Rosenblum, 1945), a fine tungsten anode wire was mounted 1·5 mm above a smooth brass plate. It was estimated that a sensitive region existed around the anode wire with a depth approximately equal to its diameter, and sparks would occur when alpha particles (but not beta or gamma) entered this region.

Connor (1951) subsequently investigated the properties of Rosenblum counters. One of the most curious findings was that the counters worked well when the active gas was air; long flat plateaux were obtained as the voltage was raised. When argon or helium were substituted, no plateaux were found. Typically, with air at one atmosphere, the counting (spark) characteristic might start at 2500 volts, rise sharply and show a plateau of 1000 volts or so. It appears that above the threshold, a quiescent corona discharge always exists. This may be indicated by a microammeter in series with the electrodes, and indeed the corona may be seen as a visible sheath around the wire. And the constantly occurring corona pulses may be observed on an oscilloscope (Prasad and Nath, 1970). The main merit of these counters is that they respond only to heavily ionizing particles— alpha particles, fragments, etc. Cosmic-rays and other background are not observed.

A number of studies have been made of the timing qualities of Rosenblum counters. Gupta and Saha (1961) estimated that the rise-time of the spark counter pulse was less than 4 nsec. On the other hand, Marinescu *et al.* (1967) found that a statistical delay occurred before breakdown with a standard deviation as large as 150 nsec. The efficiency of this counter and its application to the measurement of low alpha activities has been reported by Saro and Srkalova (1967). It has been used as a thermal neutron detector exploiting the $^{10}B(n, \alpha)^7Li$ reaction; and fast neutrons have been indicated by breakdown resulting from recoil nitrogen nuclei (Singh and Saha, 1965).

Spark counters constructed with a parallel plate geometry exhibit different characteristics. Keuffel (1949) investigated this case: a good performance with localized sparks was obtained—and spurious counts were minimized by using an argon/xylene gas mixture (380 mm Ar/6 mm xylene).

8.7 OUTWARDLY-DIRECTED AVALANCHE CHAMBER

The author has speculated on the possibility of using a proportional counter in a reverse mode, i.e. with a negative potential on the wire (Rice-Evans, 1971). In this case, the direction of avalanche development would be away from the wire, rather than towards it. Avalanches would occur only in the high-field region surrounding the cathode wire; and they are likely to be initiated by electrons released near the surface of the wire. The aim of such a configuration would be a particle detector with a high spatial precision.

The derivation of the voltage pulse induced on the wire may be obtained with a treatment similar to that outlined in Chapter 7. With a cathode wire mounted coaxially inside a cylindrical anode (Fig. 8.7.1.1), the electric field E at any point is related to the applied potential by

$$E = \frac{V}{r \ln \frac{r_2}{r_1}}.$$

A critical radius which defines the limit for secondary ionization is given by

$$r_c = r_1 \frac{V}{V_c}$$

where V_c is the threshold voltage for ionization.

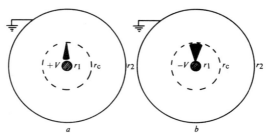

Fig. 8.7.1.1 Avalanche development in (a) conventional proportional counter, and (b) outwardly-directed avalanche chamber. Not to scale. (Rice-Evans, 1971)

Any free electron at the surface of the wire will be accelerated radially and an outgoing avalanche will be formed. The multiplication will continue until the critical radius is attained, after which the electrons will drift slowly to the cylindrical node. The positive ions will accelerate towards the cathode where, on arrival, one assumes that they are merely neutralized. On completion of the avalanche at time $t = 0$, one may assert that a shell of positive ions and electrons lies at the critical radius r_c. If the drift velocities be given by

$$u_+ = \frac{dr_+}{dt} = -k_+ E$$

$$u_- = \frac{dr_-}{dt} = k_- E$$

where k_+, k_- are the appropriate mobilities, one may show, using Green's theorem, that the growth of the cathode pulse with time is described by

$$P(t) = \frac{Ne}{C \ln \frac{r_2}{r_1}} \ln \left\{ \frac{\sqrt{2k_- Vt + r_c^2 \ln \frac{r_2}{r_1}}}{\sqrt{-2k_+ Vt + r_c^2 \ln \frac{r_2}{r_1}}} \right\}$$

where C is the chamber capacity and Ne the charge in the avalanche. The positive ions will all be collected at time

$$t_1 = \frac{(r_c^2 - r_1^2) \ln \frac{r_2}{r_1}}{2k_+ V}$$

and the electrons at

$$t_2 = \frac{(r_2^2 - r_c^2) \ln \frac{r_2}{r_1}}{2k_- V}.$$

Let the radius of the wire be 20 μm and that of the cylinder 1 cm. Suppose that the threshold voltage (V_c) is 1200 V and the operating voltage 2000 V. Then r_c is 33 μm. The avalanches will thus have a radial distance of 13 μm in which to develop. If the counter is filled with argon at atmospheric pressure, taking $k_+ = 2\cdot5$ cm^2 V^{-1} s^{-1} and $k_- = 2000$ cm^2 V^{-1} s^{-1} (Wilkinson, 1950), a fast-rising pulse is obtained. Twenty-two per cent of the maximum is reached in 10^{-10} s and 65% in 10^{-8} s. The sharp rise means that a short detectable fast pulse may be obtained with a differentiation network.

For the device to be viable as a detector, particles must produce one free electron near the wire surface. The author has estimated that there is only a small chance ($\sim 3\cdot5$%) of a minimum-ionizing particle liberating a primary electron sufficiently close to the wire, in argon at one atmosphere. On the other hand, when a particle passes through the wire, a secondary electron may be emitted from the surface. This probability, δ, is thought to be independent of particle energy at high energies. Pertinent values for δ are difficult to find. However, starting with a value $\delta = 2\cdot5$% for 19·2 GeV/c protons traversing 5 μm aluminium foils (CPS, 1969), and following trends mentioned by Bruining (1954), a very rough estimate gives 35% for the probability of an energetic proton causing the emission of a free electron at the surface of a 40 μm gold-coated tungsten wire.

Many difficulties will have to be overcome before this can become a practical device. For example, a satisfactory quenching gas would have to be found to neutralize spurious ions produced by cosmic-rays and non-intersecting particles. External quenching might also be required to cope with the avalanche ions. Field emission from rough spots on the cathode wire might prove to be a problem. The question whether a device of high-spatial resolution can be realized remains to be answered by a systematic study. Of course, the good hypothetical resolution (~ 30 μm) would be bought at the cost of a low efficiency.

8.8 STACKS OF TRANSPARENT DISCHARGE CHAMBERS

A major difficulty encountered in the operation of streamer chambers is the production of well-defined pulses of hundreds of kilovolts. Charpak, Breskin and Piuz (1972) have suggested that the wide gap of a streamer chamber could be replaced by a stack of narrow gaps formed by wire planes. The advantage would be operation at only a few kilovolts, while the capability to photograph tracks through a large volume and at almost any orientation would be unimpaired.

A novel feature in the CERN design was the careful alignment of the wires in parallel arrays, so that when viewed from infinity, the spaces between the wires presented a perfectly transparent path for the light emitted in discharges. Effectively the electrodes are rendered transparent. A large lens can be used to focus the light on to a camera objective.

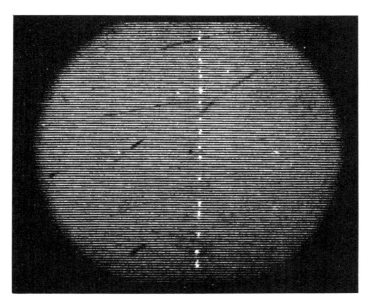

Fig. 8.8.1.1 Beta-particle track in a ten-gap chamber with a periodic wire structure. The effective transparency of the electrodes is illustrated by the illumination from behind. (Charpak *et al.*, 1972)

A small prototype 10 × 10 cm chamber was built with the gaps of 1 cm and electrodes made of 50 μm stainless-steel wires spaced every millimetre. Success as a discharge chamber was obtained with 3 kV, 1 μs pulses when the gas was a standard mixture of neon and helium; and with 6 kV pulses in the mixture plus 2% methylal $(CH_2O)_2CH_3$). Figure 8.8.1.1 shows a beta-particle track taken with a large lens, and a camera aperture of $f/5 \cdot 6$. The lattice alignment of the wires also permits photography from certain other angles (in steps of 100 mrad actually). In general, the number of discharges is about 2 per centimetre as in most streamer chambers. And yet the tracks are as bright as in most spark chambers, thus enabling one to use small camera apertures, e.g. $f/8$.

Looking to the future, stacks of discharge chambers might replace streamer chambers in large systems where multiwire proportional chambers are installed; the much smaller voltages will be tolerable! And because the light intensity is greater than that obtained with isolated streamers, digitization with vidicon recording will be possible.

CHAPTER 9

Examples of Applications of Chambers

9.1 PARTICLE PHYSICS

PARTICLE physics, of course, has been the main arena for the exploitation of spark chambers. The incessant urge to gather better data has driven experimentalists to improve and refine their instrumentation. The last decade has seen the advance from visual recording to immediate digitization, and the simultaneous development of computers has increased the scope of their potential applications.

In this chapter we shall give a few examples of particle physics experiments and then proceed to report applications in other areas.

9.1.1 Observation of two kinds of neutrino

One of the most famous experiments in particle physics was that at Brookhaven in which it was discovered that the neutrinos produced in pion decay are different from those of beta decay (Danby et al., 1962). Neutrinos resulting from

$$\pi^\pm \to \mu^\pm + (\upsilon/\bar{\upsilon})$$

were allowed to interact with the thick electrodes of an optical spark-chamber system, and a search was made for the following plausible reactions:

$$\upsilon + n \to p + e^-$$
$$\bar{\upsilon} + p \to n + e^+$$
$$\upsilon + n \to p + \mu^-$$
$$\bar{\upsilon} + p \to n + \mu^+$$

Protons of 15 GeV were deflected on to a beryllium plate to produce pions which decayed in flight. Thirteen metres of iron was placed in the path of the decay products with the purpose of absorbing the muons. Neutrinos that penetrated entered a spark-chamber system consisting effectively of forty narrow gaps defined by 2·5 cm thick aluminium electrodes. The total weight, amounting to ten tons, was enough to ensure an observable neutrino interaction rate. An interesting event consisted of a charged particle track which originated in an electrode within the fiducial volume. To this end an anticoincidence counter, which surrounded the whole chamber, comprised part of the trigger system.

In an exposure of $3 \cdot 10^{17}$ protons, 113 interesting events were recorded including thirty-four single muon events. If $v_\mu = v_e$, then a similar number of electron showers would be expected. Only six possible showers were seen, and even those were rejected on analysis. The authors concluded that there exist two kinds of neutrinos.

9.1.2 Splitting the A_2 resonance at CERN

One of the most provocative and controversial series of experiments in particle physics has concerned the splitting of the A_2 resonance (see Kienzle, 1968). The digitized wide-gap spark chambers described in Chapter 5 were incorporated into a boson spectrometer (CBS) to analyse the reaction

$$\pi^- + p \rightarrow p + X^-$$
$$(1) \quad (2) \quad (3) \quad (4)$$

where $X^- = A_2$, and $X^- \rightarrow$ decay products.

Kinematical considerations show that the missing mass M_x is given by

$$M_x^2 = (E_1 + m - E_3)^2 - p_1^2 - p_3^2 + 2p_1 p_3 \cos\theta$$

where θ refers to the recoil proton, and m is the proton mass. In the forward direction, at $\theta = 0$, $dM_x/d\theta$ vanishes, and it is sufficient just to measure p_3.

In one experiment (Benz et al., 1968) 2·6 GeV/c pions struck a 26 cm hydrogen target (Fig. 9.1.2.1). The recoil proton was detected by counter R after passing through the magnetic spectrometer. The tracks of the proton, and the decay products (e.g. $A_2 \rightarrow 3 \pi^\pm$) were recorded in spark chambers SC(1–4). The data were recorded magnetostrictively and control of the whole spectrometer system was maintained by on-line computation. Properly to identify the proton, a time-of-flight measurement was also made for the particles between counters T_2 and R.

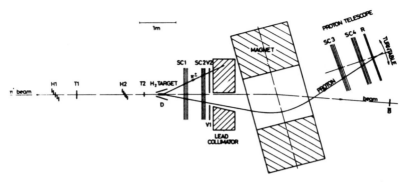

Fig. 9.1.2.1 The CERN boson spectrometer. (Benz *et al.*, 1968)

9.1.3 Colliding protons at CERN

A good example of the use of narrow-gap wire spark chambers is the CERN Intersecting Storage Ring experiment by the CERN/Utrecht/Lancaster/Manchester Collaboration (Albrow *et al.*, 1972). The purpose was to measure particle production at energies up to 53 GeV in the centre of mass.

In the I.S.R., two beams of protons circulate in opposite directions around an imperfect circle 300 m in diameter. The separation of the fixed beam paths varies around the circle and the beams intersect at eight points. The single arm spectrometer (Fig. 9.1.3.1) is capable of studying particles produced in proton collisions at one of the intersection points (I2).

A septum magnet deviates particles that lie at angles between 40 and 100 mrad, towards magnets BM1 and BM3. It is the six magnetostrictive wire chambers that measure the curved trajectories in the magnetic fields—to indicate the particle

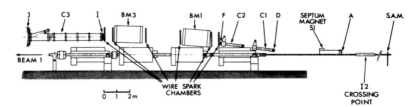

Fig. 9.1.3.1 Single-arm spectrometer for the CERN Intersecting Storage Rings. (Albrow *et al.*, 1972)

momenta. The ethylene Cerenkov counters C_1, C_2 identify kaons and pions, and C_3 (hydrogen) counts the pions only.

Production ratios for $\pi^\pm$, $K^\pm$, p and $\bar{p}$ have been successfully measured.

9.1.4 Photoproduction of hadrons at DESY

Figure 9.1.4.1 shows the experimental layout for studying photoproduction in the DESY streamer chamber (see section 6.3.4). An electron beam of up to 7·5 GeV hits a conversion target. The energies of individual photons are determined by bending the recoil electrons in a C magnet on to tagging counters (Ladage et al., 1969).

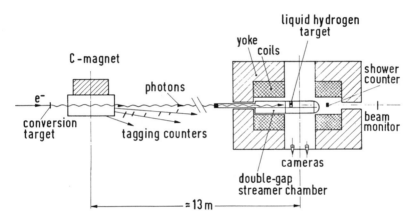

Fig. 9.1.4.1 Experimental arrangement for photoproduction at DESY. (Ladage et al., 1969)

The photons are directed on to a liquid hydrogen target that is surrounded by a 4 π plastic trigger counter. To protect the photomultiplier against the streamer light the first dynode is briefly devoltaged after a trigger. Two shower counters down stream veto events in which electron pairs are created in the target. In an experiment, 870,000 pictures were taken and 200,000 hadronic events were observed. Of these 40,000 were from the hydrogen target, and the remainder from the surrounding scintillator.

The 22 kGauss magnetic field allowed a momentum assessment of emerging particles. The resolution for a pion of 2 GeV/c with a 70 cm track is 1·5%; and for a proton of 0·4 GeV/c with a 20 cm track length, 4%—which is comparable with bubble-chamber resolutions. The measuring accuracy on film is about 6·5 μ, which corresponds with 380 μ in the chamber in the plane perpendicular to the lens axis.

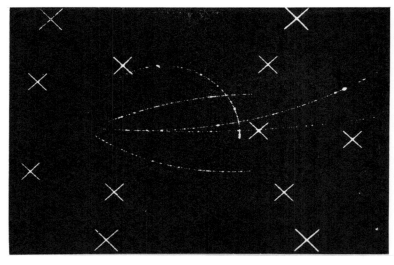

Fig. 9.1.4.2 A typical hadronic event at DESY. (Ladage *et al.*, 1969)

In a more recent experiment on electroproduction, an incident electron beam has been employed with an intensity of $2 \cdot 10^6$ electrons/sec. This intensity gives 50–80 electrons per memory time of 2 μs, which is governed by the SF_6 admixture. This time is chosen to suit the 1·2 μs delay prior to application of the field. The trigger repetition time is about ten seconds (Meyer, 1972).

9.1.5 Transition radiation detection

The advent of "super-high-energy" accelerators has called for new means for measuring particle energies. The conventional technique using Cerenkov radiation becomes impractical because, at ultrarelativistic energies, it is difficult to distinguish between particles with the same momentum but different velocities. However, a new method—based on X-ray transition radiation—shows considerable promise. Its merit is that the intensity of this radiation increases linearly with the particle energy (i.e. with the Lorentz factor, $\gamma = E/m_0 c^2$) and becomes appreciable at the super-high energies.

When a charged particle traverses an interface between two media of different dielectric properties, electromagnetic radiation is emitted at the boundary (Ginzburg and Frank, 1946). Garibian (1960) showed that at the interface between vacuum and a dielectric medium, the energy of the transition radiation emitted, per unit solid angle per unit frequency, is

$$\frac{d^2 W}{d\Omega d\omega} = \frac{e^2 \beta^2}{\pi^2 c} \frac{\sin^2 \theta \cos^2 \theta}{(1 - \beta^2 \cos^2 \theta)^2} \left| \frac{(\varepsilon - 1)(1 - \beta^2 + \beta\sqrt{\varepsilon - \sin^2 \theta})}{(\varepsilon \cos \theta + \sqrt{\varepsilon - \sin^2 \theta})(1 + \beta\sqrt{\varepsilon - \sin^2 \theta})} \right|^2$$

where θ is the angle of emission from the particle direction, ε is the complex dielectric constant and ω is the frequency of the transition radiation.

By integrating this equation over all angles and a chosen frequency range ($\Delta\omega$), and taking $\beta \to 1$, $\gamma \gg 1$, Yuan (1970) found that the emitted energy in the optical region is

$$W_{opt} = \frac{2e^2}{\pi c} (\ln 2\gamma - 3/2) \Delta\omega$$

and the total energy in the X-ray region is

$$W_{X\text{-ray}} = 2e^2 \omega_p \gamma / 3c$$

which we notice is linearly proportional to γ. The angular distribution of the transition radiation is peaked sharply in the forward direction within a cone of the order $1/\gamma$.

The results of various experiments have established the existence of transition radiation and have provided support for the theory. In particular, by studying the X-radiation emitted by 1, 2, 3, 4 GeV ($\gamma = 1000, 2000, 4000, 8000$) positrons passing through 231 aluminium foils (one mil thick, 30 mil spacing) with a Ge(Li) detector, Yuan et al. (1969) showed that the intensity of X-ray transition radiation is indeed linearly dependent on the

γ of the incident-charged particle. The finding that a 2 GeV positron emitted about 10 X-ray photons suggests that an adequate number (n) of photons would be emitted to allow satisfactory measurements on specimen particles. Furthermore, Yuan calculates that statistical fluctuations in the X-ray energy, and the X-ray number, would give an intensity resolution of $\sqrt{(2/n)}$.

The difficulties of associating (and counting) X-ray transition radiation signals from semiconductor detectors with the traversing high-energy particle has prompted Alikhanian *et al.* (1970) to propose the use of a streamer chamber to record both the particle and its transition radiation. To investigate this possibility a layered target of 1000 polyethylene films (45 μm thick, spaced at 500 μm) was placed 11 metres up beam from a streamer chamber. Any quanta produced by 2 GeV electrons, within an angle of 4×10^{-3} radians, passed down the evacuated tube and into the chamber.

The chamber was filled to 1 atmosphere with 70% neon/30% helium plus a 10% addition of xenon to give a high detection efficiency for X-rays. Iodine vapour (10^{-2} Torr) was also introduced for proper functioning. The chamber was operated in the projection mode, i.e. photographs were taken through the electrode along the electric field. The arrangement enabled the particles to be detected with 100% efficiency, while the high spatial and angular resolution of the chamber allows the counting of electrons liberated by transition photons over all angles. Figure 9.1.5.1 shows a typical chamber photograph in which a primary particle is accompanied by two transition photons.

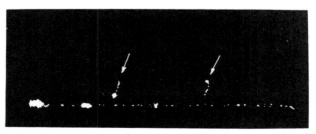

Fig. 9.1.5.1 Streamer-chamber photograph showing primary track and photo-electrons produced by two transition photons. (Alikhanian *et al.*, 1970)

After an analysis of many events, the authors concluded that the number of quanta increased linearly with electron energy in the range 1–2·4 GeV. The background due to delta rays and bremsstrahlung proved to be relatively small. With the layered target, a detection efficiency of 67% was obtained at 2 GeV. When this target was replaced with a plastic foam target it was found that the efficiency of electron detection by transition radiation rose to 86%. These preliminary studies promise success at super high energies.

A very promising technique is to employ a sandwich configuration of thin proportional chambers and foam transition radiators. Even with large numbers of sandwich elements, the overall energy loss may be negligible (~ 100 keV for 10 GeV particles, say) if thin-walled Charpak chambers are used. The chambers (filled with argon, or better still, xenon) will detect the X-rays (1–20 keV) with high efficiency; and they can cover large areas and angles.

A major problem, of course, is that the ionization due to the traversing particle is registered in addition to that resulting from the transition X-ray. Large output pulses should indicate X-rays. However, a complication is presented by the Landau tail (see section 7.5.7)—the occasional large energy deposits call for a statistical approach with a cascade of proportional chambers.

Yuan *et al.* have considered applying the geometric mean technique to the outputs of twenty-five proportional chambers. A shift of the mean to higher values indicates the presence of X-rays, and the magnitude of the shift should be correlated with γ, the particle energy. (See Fig. 9.1.5.2.)

9.1.6 Search for fractionally charged particles

The mythical quark has tantalized romantics in recent years; heroic quests have been made to discover the particle predicted to have a charge of $e/3$ or $2e/3$. Oysters have been tasted, diamagnetic specks levitated and optical spectra peered at (Massam, 1968). However, most methods have relied on the effect of the reduced charge on the specific ionization; dE/dx for a particle traversing a plastic scintillator, say, should be proportional to the square of the charge.

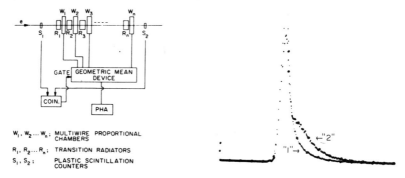

Fig. 9.1.5.2 (*Left*) Assembly for detection of transition radiation. (*Right*) Pulse-height distribution of 10 GeV electrons in a single proportional chamber, "1" for ionization loss only, "2" with additional contribution of transition radiation created in a foam radiator. (Yuan et al., 1972)

An advantage to be gained by using spark chambers in the apparatus is that the record of the tracks can remove ambiguities due to showers, etc.; and the absence of tracks can indicate that low-ionization signals might be due to particles passing through light guides. An early approach was to reduce the efficiency of conventional multigap spark chambers by adding electronegative impurity gases; the number of gaps sparking then gives a measure of the ionizing power of the traversing particle (Buhler-Broglin et al., 1966). With a suitable delay (t) before application of the high-voltage pulse, the number of ions produced in the gap (N) is related to the efficiency (ε) by

$$\varepsilon = 1 - \exp[-N \exp(-t/T)]$$

if it be assumed that the loss of ions due to the impurity is exponential with decay time T.

A good example of a systematic search was that by Allaby et al., 1969, who employed the CERN avalanche chamber (see section 6.6) to look for negative quarks produced in an internal synchrotron target, by pair production

$$N + N \to N + N + Q + \bar{Q}$$

or by dissociation

$$N + N \to N + Q + Q + Q$$

A secondary beam channel was designed to accept negative particles produced, in the forward direction, of 1·2 times the proton momentum—i.e. to accept either 10·9 GeV/c, $q = -1/3$ or 21·7 GeV/c, $q = -2/3$ quarks. The detection system comprised six plastic trigger counters sensitive to low ionization, six plastic pulse-height counters, and the avalanche chamber. The $10 \times 10 \times 11$ cm³ chamber, when coupled to an image intensifier and camera, could record individual primary avalanches along the track. Therefore the tracks of any likely candidates could be assayed and the charge determined from the number of avalanches.

In practice, due to the finite size of a streak on the film, the number (N) of ionizations per unit length must be determined from the number (n) of observed avalanches taken into account, by the formula

$$N = n/(l - n\Delta)$$

where Δ is the interval following a counted avalanche, in which for counting purposes all avalanches are ignored; and l is the track length.

With a distribution function

$$P(N, n) = \frac{e^{-(l-n\Delta)N}\{(l - n\Delta)N\}^n}{n!} + \frac{N}{(n-1)!}$$
$$\int_{l-\Delta}^{l} e^{-(x-(n-1)\Delta)N}\{(x - (n-1)\Delta)N\}^{n-1} \, dx$$

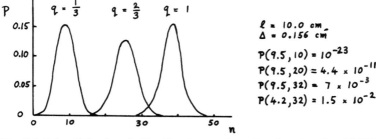

Fig. 9.1.6.1 Calculated distributions of avalanches in the CERN 10 cm chamber corresponding to quark candidates. The average ionization (N) for a singly charged particle was found to be 9·5/cm; and it follows that the probability of observing only ten avalanches (n) in the chamber was 10^{-23}. (Allaby et al., 1969)

the expected distributions of Fig. 9.1.6.1, corresponding to charges $e/3$, $2e/3$ and e, are seen to be well separated. In the experiment care had to be taken to eliminate erroneous tracks due to voltage pulses having too small an amplitude or duration (i.e. monitoring was necessary), and due to electrons lost by attachment in the gas, or to the walls of the chamber, in the 350 ns delay before application of the pulse. Disappointingly, the CERN authors found that the measured distribution corresponded only to the charge e, and concluded no quarks were seen.

9.2 LOW-ENERGY PHYSICS

9.2.1 Alpha rays in a streamer chamber

A number of attempts to use streamer chambers in the observation of alpha particles have been reported. So many specific primary ions are created along the path of an alpha particle that one would not expect any difficulty in observing streamers —rather, it is the short range that proves awkward. Nevertheless, alpha tracks from radioactive sources have been determined. And recoil nuclei from $n - \alpha$ collisions have been measured in a helium-filled chamber to detect neutrons ($\geqslant 5$ MeV) and analyse their degree of polarization (Giacomich and Lagonegro, 1965).

A problem encountered in the study of low-energy particles arises from the need to trigger the chamber. The normal procedures of high-energy physics, involving complementary exterior triggering scintillation counters, are inapplicable because the particles do not have the energy necessary to emerge from the chamber and penetrate a counter. A sensible approach is to arrange for the means of triggering to be incorporated in the chamber. Thus Cavalleri *et al.* (1963) obtained trigger pulses by detecting the primary gas scintillations with a photomultiplier. Similarly, Yasumi *et al.* (1964) detected the Cerenkov light radiated in the chamber gas by the traversing particle. Giacomich and Lagonegro (1965) coated the floor of their chamber with a thin lining of ZnS(Ag), covered it with a thin sheet of aluminium, and detected the scintillations therein.

An arrangement that has been found satisfactory for cases in which a radioactive specimen is introduced into the chamber volume, incorporates plastic phosphor as the bottom side of the chamber (Rice-Evans and Mishra, 1971). In this design a 22 × 22 × 1·5 cm slab of NE102A phosphor, which seals the chamber volume with an "O" ring, is coupled to a photomultiplier in the conventional fashion via a triangular light guide (Fig. 9.2.1.1). Any low-energy particles striking this slab are detected and trigger the chamber. Any unwanted triggers that might arise from cosmic rays are prevented by placing a

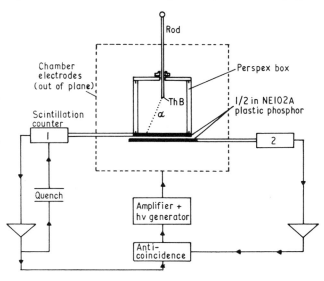

Fig. 9.2.1.1 Streamer-chamber arrangement for recording tracks of low-energy charged particles. (Rice-Evans and Mishra, 1971)

second piece of phosphor immediately beneath the first. Connected in anticoincidence, a trigger results only if downward-moving particles are stopped in the upper phosphor.

One consequence of this open design was the need after triggering, to prevent the ruin of the photomultiplier by light emission from the subsequent streamer discharge. Accordingly, a monostable circuit was built which would deliver a negative 250 V, 8 ms, pulse to the first dynode of the photomultiplier, just before the commencement of the discharge. This "quenching" circuit consists of valves which, in our experience, are more reliable than transistors in the neighbourhood of streamer chambers. The additional precaution was also taken of using high (1·6 MΩ) resistances in the thirteen-stage dynode chain. This protects the anode of the photomultiplier in the event of an undesirably large pulse, by effectively reducing the consequent anode current.

In the diagram the electric field is perpendicular to the plane of the paper. The photograph shows typical alpha-particle tracks from a Thorium B source, taken with a camera viewing the chamber through a transparent electrode comprised of fine

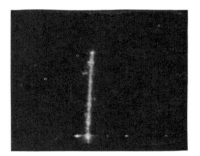 

Fig. 9.2.1.2 Alpha-particle tracks in a streamer chamber. (Rice-Evans and Mishra, 1971)

wires spaced at 2 mm. Unlike the case of a minimum-ionizing particle in which the track is depicted by a line of individual isolated streamers, the alpha-particle track is diffuse and relatively broad. This is no doubt due to the much bigger primary ionization—thousands of ions per centimetre—created by the alpha particle. Beta particles give a finer streamer trail.

The 6·1 and 8·6 MeV alpha from Thorium B have ranges of 12 and 17 cm in the standard (70/30) neon/helium mixture at one atmosphere. At lower energies, a limit is imposed by the short range and the need to reserve some energy for the scintillation process. One approach would be to reduce the gas pressure (experience shows that streamer chambers will operate at 0·5 atmospheres). This, however, would result in broader tracks because of the enhanced diffusion.

9.2.2 Rare events—search for double beta decay

The streamer chamber is an instrument of unique sensitivity and discrimination. These qualities derive from the fact that the triggering system may be devised to respond selectively to rare events of interest. Furthermore, the resulting photographs may be studied at leisure and spurious background events discarded. Recognizing this, Mrs. C. S. Wu (Bardin *et al.*, 1967; Ullman *et al.*, 1968) performed an experiment that placed a lower limit on the lifetime of the lepton non-conserving neutrinoless mode of double beta decay in ^{48}Ca of $1 \cdot 6 \times 10^{21}$ years.

In the experiment, ten grams of ^{48}CaF$_2$ source, sandwiched between two thin 50 cm diameter aluminium foils, formed the central electrode of a double (4 cm) gap chamber (Fig. 9.2.2.1). The outer electrodes, 70% transparent grids of 8 μm nickel, were enclosed with 20 μm mylar gas windows. Beyond the windows on either side were placed sixteen scintillation counters —comprised of 2 cm thick, wedge-shaped plastic phosphor segments to cover the whole 50 cm circle. A magnet coil produced 400 Gauss within the chamber. Each gap was viewed in the projection mode by a camera looking through the scintillators. To reduce background, the whole apparatus was placed 2000 feet below ground level.

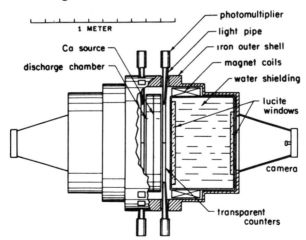

Fig. 9.2.2.1 Streamer chamber for double beta decay. (Bardin *et al.*, 1967)

The system triggered if the scintillators produced two coincident pulses with appropriate pulse heights, i.e. within a range chosen to indicate possible ββ events with a maximum sum energy of 4·24 MeV. In 1100 hours of operation, the event rate was about 40 per hour. Most were single electrons passing through the chamber. Subsequent study of the film, requiring that the two tracks must come from the same point in the source, and also show the expected curvature in the magnetic field, led the authors to reject all but four events. This indicated a lower limit to the lifetime, although the four observed events might be due to contaminant. However, the experiment certainly supported the postulate that leptonic number is conserved.

9.2.3 Fermi surfaces—positron annihilation photons

When a positron annihilates in a crystal, two 511 keV photons are emitted from the site in opposite directions. Any departure from 180° gives an indication of the momentum of the electron participating in the annihilation; the positron motion is assumed to be negligible. The electron momentum distribution corresponds to positron annihilation with either the conduction electrons or the core electrons—but it is possible to separate the two contributions and thus arrive at the Fermi surface.

The momentum components are given by

$$p_x = m_0 c \theta_x, \; p_y = m_0 c \theta_y$$

where θ_x and θ_y are the projected angles of deviation. Conventional experiments, employing scintillation counters behind collimating slits in lead blocks, have the handicap that only one component is measured (Stewart and Roellig, 1967). Howells and Osmon (1971, 1972), however, have incorporated a spark chamber for the detection of one of the gamma rays—with the aim of measuring θ_x and θ_y simultaneously.

Their apparatus (Fig. 9.2.3.1) consisted of two arms: one a NaI(Tl) scintillation counter mounted behind a pinhole collimator and the other a 60 gap, 10 × 10 cm spark chamber with 0·12 mm aluminium electrodes spaced at 2 mm. The spark chamber was triggered whenever a gamma ray of the correct energy was detected in the NaI crystal. The high pulse rates and large chamber capacitance (5000 pF) called for a stripline pulser (previously developed by the same Westfield College group), (Howells *et al.*, 1970) coupled with a thyratron; the 1 kV, 40 ns rise-time pulses were sufficient to respond to the ionization produced by any Compton electrons ejected from one of the electrodes. The chamber detected 511 keV gamma rays with an overall sufficiency of 2%. The 60 gaps were scanned with a television camera operated on a 60-line raster; and digitized spark locations were found with the digitron technique of counting oscillator pulses stored in a scaler.

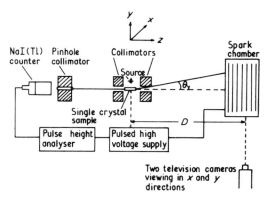

Fig. 9.2.3.1 Angular correlation apparatus for positron annihilation studies. (Howells and Osmon, 1971)

In an experiment on annihilation in a single copper crystal, the chamber was placed 4 m from the crystal—a distance determined by the need to achieve an angular resolution of 0·5 mrad over a range of 0–15 mrad. A delay of 620 ns between the arrival of the gamma and the delivery of the high-voltage pulse imposed a limit on the source strengths that could be used—due to random events. It was found that a 5 Ci copper 64 source ($T_{\frac{1}{2}} = 12\cdot6$ hours) gave 50,000 events; on which an unsophisticated analysis just indicated the "necks" of the copper Fermi surface. This new approach to electronic structure studies has scope for refinement and therefore bodes well.

9.3 COSMIC RAYS

9.3.1 Spark chambers for cosmic rays

The Michigan experiment (Jones *et al.*, 1972) performed at Echo Lake, Colorado (3230 m) is a good illustration of the use of spark chambers in cosmic-ray physics. The purpose was to study proton–proton interactions in the energy range 100–1000 GeV.

The apparatus (Fig. 9.3.1.1) consisted of a liquid hydrogen

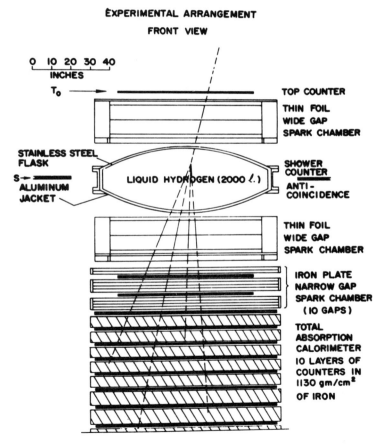

Fig. 9.3.1.1 Michigan cosmic ray apparatus (Jones *et al.*, 1972)

target and an assembly of wide- and narrow-gap optical spark chambers. The wide-gap chambers placed above and below the target were 2×2 m² in area and each gap was 20 cm. The electrodes were 51 μm hardened aluminium and the side walls plate glass. Each two-gap chamber was pulsed with an eight-stage Marx generator (3900 pf/stage; 15 kV/stage). Ten narrow (1·9 cm) gap chambers sandwiched between copper-clad iron plates formed part of an ionization calorimeter. All the chambers were photographed with 90° stereo, 65:1 demagnification, $f/10$ aperture and Kodak tri-X film.

Conclusions were drawn on the p–p cross-sections and the multiplicity distribution of charged prongs.

9.3.2 Gamma-ray astronomy

The prospect of measuring celestial gamma rays has tantalized astronomers in the last decade. Searches might be expected to indicate point sources in the Universe, and to provide checks on postulated astrophysical mechanisms, e.g. neutral π mesons produced in the interaction of cosmic rays with interstellar matter will decay into two gamma rays, each of about 70 MeV in the rest frame. Bremsstrahlung, synchrotron radiation and the inverse Compton effect are other likely processes. The necessity to escape the Earth's atmosphere, implying balloon or satellite-launched apparatus, has prompted the use of relatively light spark-chamber systems. Furthermore, the imaging properties of spark chambers should enable the background to be rejected—for the problem is one of distinguishing an intensity of, say, 10^{-6} gamma rays/cm²/sec from the charged particle flux of about 1/cm²/sec.

Useful reviews of gamma-ray astronomy have been written by Garmire and Kraushaar (1965) and Fazio (1970). Several telescopes have been described, e.g. Board et al. (1968), Dean et al. (1968) and Niel et al. (1969). For our example we shall take the Goddard Space Flight Centre balloon-borne apparatus used to seek gamma rays from the galactic centre, Virgo and the Crab Nebula (Fichtel et al., 1972).

High-energy gamma rays (30–500 MeV) interact with matter predominantly by pair production. The task is therefore to allow gamma rays to strike suitable foils and to record events in which two spark-chamber tracks (e^+, e^-) appear to emanate from a single point. The Goddard design (Fig. 9.3.2.1) consisted of a stack of thirty-two narrow-gap (50 × 50 cm) modules; each module being composed of two orthogonal sets of 400 parallel wires forming wire grids spaced at 4 mm. Stainless-steel plates with a thickness of 0·03 radiation lengths lay between the

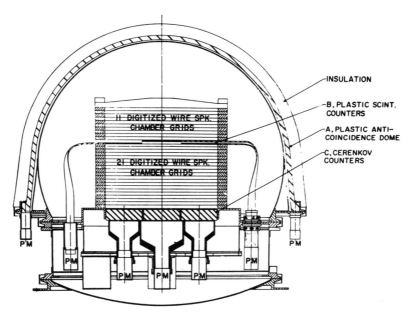

Fig. 9.3.2.1 Goddard-digitized spark chamber gamma-ray telescope. (Fichtel *et al.*, 1972)

eleven upper modules to convert the gamma rays into pairs. Plates of 0·02 radiation lengths lay between the lower modules to give information on the energy of individual positrons and electrons by Coulomb scattering (Ross *et al.*, 1969).

The spark-chamber triggering logic was chosen to accept gamma rays and reject other events. A coincidence signal was required from one of the nine internal particle telescopes (each consisting of a scintillator and a perspex Cerenkov counter). This ensured that a trigger resulted when a neutral particle was converted into one or more downward-moving relativistic-charged particles. Traversing primary-charged particles were rejected by anticoincidence signals produced in the surrounding plastic scintillator dome. The latter, 1.2 m in diameter, was viewed by eighteen photomultipliers. The data were recorded digitally, with ferrite cores threading each wire, and telemetered to the ground, together with the three axis magnetometer readings which gave the balloon position, etc.

In the first flights no evidence was discovered for radiation from the sources M87, 3C273 or the Crab Nebula, although an excess flux above atmospheric background was found for gamma rays exceeding 100 MeV coming from the galactic centre.

9.4 STUDY OF ELECTRICAL DISCHARGES

9.4.1 Electrical breakdown along an ionized trail

A good example of the application of spark chambers in the systematic investigation of gas discharges is the experiment by Caris *et al.* (1968). Figure 9.4.1.1 shows their treble gap chamber. Each gap was 4·5 cm. The chamber could be rotated in a beam of 1 GeV pions. The triggering system applied a rectangular high-voltage pulse to the chamber 400 ns after a pion had passed along the line defined by the two scintillation counters. The chamber was built as a matched element in a transmission line terminated by its characteristic impedance of 50 Ω. When the pressurized spark gap was triggered, it discharged a length of charged cable. A pulse, of length twice the electrical length of the charged cable, was then transmitted with an amplitude equal to half the charging voltage (40–100 kV). A controlled flow of henogal (70% neon/30% helium) passed through the chamber at a defined pressure (680 or 689 Torr).

Measurements were made on the time required for spark formation—as a function of the field strength and the track orientation. Oscilloscope traces recording the voltage across the termination resistor were photographed. The moment of spark breakdown corresponded with the collapse of the high-voltage pulse—for the spark channel provided a short circuit. One interesting finding was the way the breakdown times increased as a function of track angle.

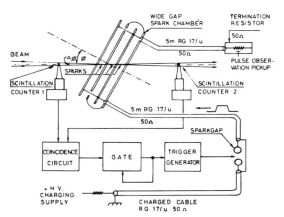

Fig. 9.4.1.1 Investigation of the gas discharge along an ionized trail. (Caris et al., 1968)

To determine the rate of growth of light emission in a discharge, the apparatus in Fig. 9.4.1.1 was modified to provide a high-voltage pulse with a continuously variable length. This was achieved by locating a spark gap at each end of the charged cable. One spark gap was connected to the chamber, and the other directly to a termination resistor. The length of the high-voltage pulse was varied by changing the relative delay in the firing of the two spark gaps. The light emitted in the chamber was imaged with a lens on to a diffuser screen placed before the cathode of a photomultiplier. The range of detection was extended by using neutral filters. The current pulse was integrated and fed into a 100-channel pulse-height analyser. The curves of Fig. 5.2.1.3 were obtained. One result of the data on the temporal growth of a spark was that it was possible to extend the known range of values of αu to higher electric fields. The assumption is made that light emission is proportional to electron number. The term αu is important in electron multiplication, $n = n_0 \exp(\alpha u t)$. Here u is the electron drift velocity.

9.4.2 Streamer formation in hydrogen

The prospect of a new epoch in gas-discharge physics is presented by the recently developed technology of streamer chambers.

The ability to apply reproducible and measurable pulses of desired high amplitude, short duration, and shape to a discharge chamber means that the various stages in the avalanche/ streamer growth may be studied with a new precision. Also, by concentrating on streamers initiated by single electrons in the middle of a large gas volume one might hope to preclude secondary effects such as electron emission from the cathode and thus investigate the gas-multiplication mechanism in isolation.

Before a complete theory of streamer formation can be proposed an empirical onslaught will have to be made on a large number of gases; and correlations searched for between the temporal/spatial formation and the gas characteristics, i.e. the ionization and excitation processes; the energies and lifetimes of levels, etc. A sustained attack on the noble gases is now a definite possibility—indeed in the course of this book we have frequently encountered contributions, e.g. the neon streamers shown in Fig. 1.4.4.2. However, in this section we shall allude to the difficult case of hydrogen which may lie just within the limits of what is presently practical.

Although the CERN group (Schmied *et al.*, 1972) encountered difficulties in devising a hydrogen streamer chamber (see section 6.5.4), their effort did throw some light on the streamer mechanism in hydrogen. One discovery was the extraordinarily rapid longitudinal growth of the streamers. A consequence was that the discharge could not be satisfactorily limited; the 4 cm interelectrode gap was inevitably traversed and secondary effects immediately came into play at the cathode. Bayle and Schmied (1972) referred to the level diagram of Fig. 9.4.2.1 and pointed out that this could be explained by the lifetimes of the predominant transitions $B \to A$, $C \to A$ (10–13 eV) which are only 0·8 and 0·6 ns respectively, and by the high transparency of hydrogen to ultra-violet (absorption is only perceptible for wavelengths below 1115Å).

Since the visible transitions ($C \to B$) represent only 1·6% of the luminous emission, compared to 63% in the far ultra-violet, the low absorption of ultra-violet photons must result in the ionization fronts advancing with almost the speed of light. For some practical purposes, if a mixture can be tolerated, organic vapours like alcohol may be added to absorb the ultra-violet.

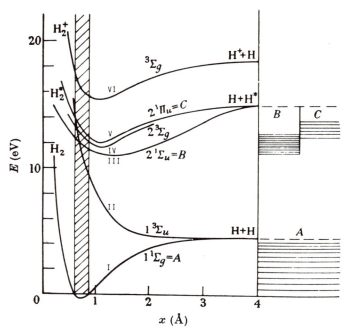

Fig. 9.4.2.1 Energy levels for the hydrogen molecule H_2. (Corrigan and von Engel, 1958)

However, for pure hydrogen, Bayle and Schmied concluded that satisfactory streamers would require high-voltage pulses of only 1 ns duration!

9.4.3 The ringing of Lichtenberg figures

In 1777, the diarist, courtier, physicist and aphorist Georg Christoph Lichtenberg published in Göttingen his discovery that when an electrical discharge occurs in the proximity of a solid dielectric, gorgeous star-like patterns may be manifested on its surface.

> Now it so happened that the lid, suspended from the ceiling, was lifted off the base for a time, so that dust could fall upon the base itself; and now it did not settle regularly, as before, on the lid, but to my great delight formed little stars in several

places, which were at first unclear and difficult to distinguish, but later, when I deliberately strewed on the dust more thickly, became clear and beautiful, resembling in places a design in relief. Countless little stars showed up here and there, whole Milky Ways and greater Suns; there were arches, indistinct on the concave side but on the convex decorated with rays; then neat little branches, not unlike those which the frost forms on window panes; little clouds of many shapes and shadings; and finally many figures of special shapes.*

And in the intervening two centuries, the phenomenon of Lichtenberg figures has been studied in some detail (e.g. see Merrill and von Hippel, 1939 and Morris, 1951).

In the streamer chamber, unlike the spark chamber, a dielectric is interposed between the electrodes and the sensitive gas volume. Electrons and ions from an electrical discharge in the gas will tend on arrival at the dielectric surface to move to cancel the polarization charge. Resulting radially-spoked Lichtenberg figures have frequently been photographed in the Bedford College streamer chamber.

The curious rings of Fig. 9.4.3.1 were discovered when the conventional pulse-limiting shunting spark gap was replaced by a Lecher's wires arrangement (Rice-Evans and Hassairi, 1972). Initially, no care was taken to absorb the voltage reflected from the shorted termination, so it is probable that the field in the chamber oscillated, swinging from one direction to the other (Fig. 6.3.1.5).

The rings, which were visible to the eye, showed the red colour characteristic of the track discharge in the neon/helium. Gaseous ions were thus responsible for the luminosity. In an oscillating field the avalanches, initiated by the specific primary ionization of the cosmic ray, would be expected to grow, to and fro, along the line of the track. The discharge would then constitute an oscillating plasma and would give rise to a fluctuating magnetic field with cylindrical symmetry.

The figures must indicate the form of the discharge that results from electrons moving radially over the surface of the perspex. A relevant fact is that a spark discharge on a dielectric

* Translation taken from J. P. Stern: *Lichtenberg—A Doctrine of Scattered Occasions*. (Thames and Hudson, London, 1963).

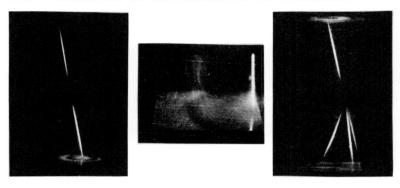

Fig. 9.4.3.1 Lichtenberg figures associated with tracks of cosmic particles in the streamer chamber. (Rice-Evans and Hassairi, 1972)

surface is known to have a greater luminosity than a free discharge in the same gas (Daniel, 1965). The dark rings may be explained by the electrons moving away from the dielectric; perhaps fluctuating polarization charges or alternating magnetic fields produce successive damping and reignition of the radial discharge. A proper interpretation must await measurements correlated with an accurate assessment of the pulsed field. A systematic study of the rings might then provide a new insight into the behaviour of electrons and ions near dielectric surfaces.

9.5 APPLICATIONS IN MEDICINE AND BIOLOGY

9.5.1 Applications of isotopes in medical diagnosis

In medical and physiological studies, radioactive isotopes are commonly used for diagnostic purposes. For example, diseased parts of organs may be located by studying the uptake of activity. If a patient ingests the X-ray emitter I^{125}, a determination of the spatial distribution of, say, 20 μC absorbed in the thyroid gland may indicate goiter. Such studies may be made by scanning with a scintillation counter or, in recent times, by coupling a mosaic of scintillation crystals to a computer. The focusing of X- and gamma-rays is not possible and so a collimator (usually a block of lead with many small diameter parallel cylindrical holes cut in it) is used to select just parallel rays emerging from the source area.

The high-spatial resolution of the spark chamber when coupled with a convenient read-out system, has prompted attempts to develop a spark-chamber camera for clinical use. If photons are to be detected, they must be made to initiate the chamber action. The efficiency of the photoelectric process increases with diminishing photon energy so that whereas an external converter (e.g. 1 mm Pb foil) might be necessary for gamma energies, in the X-ray region the chamber gas and electrodes may suffice.

Apart from gamma scanning, many biological investigations rely on determining the distribution of beta-emitting isotopes, e.g. in chromatography and electrophoresis. Tritium ($^{3}_{1}$H), carbon-14 and phosphorus-32 are important in this connection. In these applications, the introduction of the beta rays to the sensitive gas volume presents a difficulty. And in both beta and gamma applications, triggering can be a problem.

9.5.2 The spark chamber approach

One of the first chambers developed for gamma-ray imagery was that by Lansiart and Kellershohn (1966). In essence, their 20 cm diameter chamber was a triode comprised of a cathode, a central grid and a glass anode; filled with xenon plus 3 cm of methylal for quenching (Fig. 9.5.2.1). Low-energy gammas pass through the aluminium cathode and produce photoelectrons in

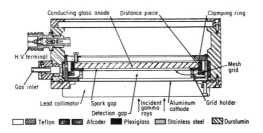

Fig. 9.5.2.1 Medical spark chamber by Lansiart and Kellershohn (1966).

the xenon within the large (20 mm) cathode-grid gap. These are accelerated through the grid by a P.D. of 1000 volts into the 3 mm anode-grid gap where they cause spark breakdown as a result of 7300 volts across this gap. (Note: ancillary triggering is not used here.) The sparks are photographed through the tin oxide-coated glass anode, with the camera shutter remaining open for the acquisition of many spark images. Figure 9.5.2.2 shows the gall bladder of a dorsally recumbent subject after

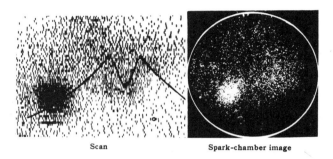

Scan   Spark-chamber image

Fig. 9.5.2.2 Comparison of images of a gall bladder impregnated with iodine 125, taken with a scintiscanner and a spark chamber. (Lansiart and Kellershohn, 1966)

intravenous injection of 500 μC rose bengal labelled with I^{125}. Mercury-197, which emits 80 keV gamma and 70 keV X-rays has also been employed—e.g. a subject's kidney was imaged after an intravenous injection of 500 μC of $Hg^{197}Cl_2$; and the spleen after Hg^{197} labelling with bromo-1-mercuri-2-hydroxypropane. Horwitz et al. (1965) have made similar studies, including the brain.

Pullan et al. (1966) developed a crossed-wire spark chamber for assessing the distribution of beta-emitting isotopes—e.g. carbon-14 and tritium—on two-dimensional chromatography and electrophoresis papers. In their 8 × 8 in. chamber, the chromatography plate was introduced into the chamber to lie just below the cathode wire plane which itself was separated from the anode plane by 2·5 mm. A continuous flow of 90% argon, 10% methane was maintained, and a steady voltage (< 5000 volts) was applied to the gap through a high resistance. The sparks occur at positions matching the active spots on the plate because of the short range of the beta particles. Each spark discharged the capacity of the chamber, and the voltage has to recover before the next spark. The counting rate was 20 cps. A magnetic core read-out from the wire planes enabled the spark positions to be recorded.

The early medical chambers so far described have been handicapped by the application of steady sparking voltages across the electrodes; a consequence is that spurious sparking is almost impossible to prevent. A better approach, using the instrumentation of contemporary particle physics, has been adopted by Kaufmann *et al.* in California (1971). They applied pulsed voltages to crossed wire planes and used magnetostrictive read-out to determine the wires participating in the spark.

The crucial triggering system is shown in Fig. 9.5.2.3; a steady high voltage on the 0·08 mm steel wires of the central plane causes electron multiplication and a proportional signal is fed to the voltage-sensitive preamplifier. As a result a high-voltage pulse is placed on the spark chamber via a series spark gap. The preamplifier is protected from sparks by the diodes. With a wire plane spacing of 1 cm; 90% neon, 10% helium,

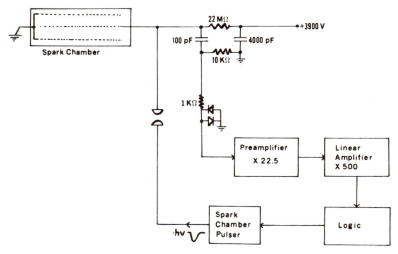

Fig. 9.5.2.3 Spark-chamber triggering by gas-multiplication. (Kaufman *et al.*, 1971)

5 cm alcohol in the chamber; a detection efficiency of 100% was achieved with 3800 volts d.c. on the central wires.

Whole bodies impregnated with high-energy gamma-emitting isotopes (e.g. cobalt-60: 1·17 and 1·32 MeV) could be viewed when a 0·06 in. lead converter was incorporated as one side of the 45 × 45 cm chamber. A 15 cm thick lead collimator with 5 mm diameter holes and 3 mm septa was suggested. After magnetostrictive collection of data, a computer analysis gives a construction of the image.

Some of the problems met in nuclear medicine—when isotopes are used—are similar to those described in section 9.2. For example, the work of Howells and Osmon (1971) on the detection of 511 keV photons from positron annihilation could be pertinent to clinical measurements.

9.5.3 Proportional chambers for nuclear medicine

The multiwire proportional chamber is more promising than the spark chamber for many medical purposes. This is because of its fast response, its small dead-time and the fact that it does not need to be triggered. Gases with a higher atomic number than neon may be used, with the advantage that the greater photoelectric absorption will increase the efficiency for stopping photons. Xenon is very satisfactory, and this element also offers the prospect of use at high pressures, and as a liquid.

Kaufman *et al.* (1972) have described a collimated wire chamber imaging system for recording characteristic X-rays emitted from organs after excitation with gamma rays. Specifically, they used 60 keV gammas from the Am241 to excite the 26·8 keV X-rays of natural non-radioactive iodine taken up by the patient's thyroid. The chamber consisted of three wire planes. The outer cathode planes, whose wires were at 90° to each other, sandwiched a central anode plane of 30 μm wires. With a 90% Xe/10% CO_2 gas mixture, X-rays had a 13% chance of liberating in the gas an electron that would result in an avalanche at a central wire.

The x and y coordinates of the avalanche were determined with the pulses induced in the neighbouring cathode wires by the movement of the avalanche ions. Instead of attaching amplifiers to each wire, the location of the appropriate wires was achieved with delay lines (Grove *et al.*, 1972). At each cathode a line lay orthogonally over the wires in such a way that a pulse on any wire was transmitted to a corresponding position on the length of delay line by capacitative coupling. The times taken for the pulses to travel to the ends of the delay line, measured with zero-crossing discriminators and time-to-pulse height converters, indicate the location. The 11×13 cm^2 prototype chamber gave a 2 mm spatial accuracy and could be operated at rates of 10^5/sec with a rejection of 30%.

Borkowski and Kopp (1972) have also described a proportional chamber photon camera. One novel feature was the addition of a drift region to improve the detection efficiency. A uniform electric field is maintained in this region so that any electrons liberated drift towards, and along a path perpendicular to the plane of, the anode wires. The read-out was from the two orthogonal cathode wire planes. By using high-resistance wires, the cathodes essentially form distributed RC lines. Any avalanche position, i.e. where the pulse was induced, along a wire, could then be obtained by measuring the rise-time of the pulse on its arrival at the end of the wire. The chamber had a resolution of 1 mm.

The prospects of the liquid xenon chamber (see section 8.5.3) as a radioisotope camera have been considered by Zaklad *et al.* (1972). With the operation in the ionization mode the authors concluded that a camera was feasible with a resolution better than 1 mm and a counting capability higher than 10^6/sec.

The advantage of a fast response is that dynamic studies become possible. For example, estimates of regional ventilatory function, and the determination of haemodynamic parameters such as cardiac output. With suitable analysis and output electronics the multiwire chamber is likely to surpass conventional imaging cameras.

9.6 ARCHAEOLOGY

9.6.1 Chephren's pyramid

Three pyramids, built in about 2500 B.C., stand at Giza, a few miles from Cairo. They are the Great Pyramid of Cheops (145 m high), the Second Pyramid of Chephren (145 m) and the Third Pyramid of Mycerinus (Fakhry, 1961). Cheops' Pyramid shows great complexity in construction, with many internal passages, galleries and chambers. In contrast only one burial-chamber has been discovered in Chephren's Pyramid.

In 1965 Luis Alvarez suggested that a search might be made for possible cavities in the limestone structure by operating spark chambers in the known burial-chamber (Fig. 9.6.1.1) and

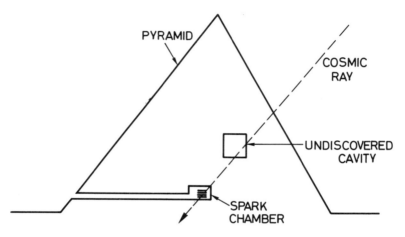

Fig. 9.6.1.1 Search for hidden chambers in Chephren's Pyramid. (Alvarez et al., 1970)

looking for anomalies in the angular distribution of cosmic-ray muons; an enhanced intensity in one direction might indicate a hollow chamber above. Accordingly, a joint Egyptian/American team installed a spark-chamber telescope (Alvarez et al., 1970).

The apparatus consisted of two narrow-gap wire-plane chambers (2×2 m) separated by 30 cm; arranged between three layers of plastic scintillator. Triple coincidences in the scintillators signalled a passing muon and triggered the chambers; and the x and y coordinate of each spark were recorded magnetostrictively (El Bedewi et al., 1972).

The experiment would have been impractical if the scattering of high-energy muons in limestone had been much more than expected; but the success of the telescope in recording the ridges of the pyramid removed all doubts. It was estimated that a muon's path in limestone could be determined to an accuracy of 1 metre in 100. In spite of several premature indications the final analysis on two million tracks disappointingly revealed no such delight as a King's chamber or a Grand Gallery. They must lie to the side?!

CHAPTER 10

Drift Chambers

10.1 THE BASIC IDEA

EXCELLENT prospects for drift chambers were predicted at the Frascati conference in May 1973. Although the concept had been advanced by Charpak and his group in 1969, general acceptance had to wait for the reports of pioneering groups with operating experience. At the conference, results from CERN, Heidelberg and Saclay were presented, huge chambers built at Harvard were described, and very optimistic conclusions were drawn.

The essential idea may be seen in Fig. 10.1.0.1. A uniform electric field is maintained in most of the chamber so that ionization electrons liberated at the particle trajectory will drift with a constant average velocity towards the proportional counter. The time that elapses between the particle's passage and the avalanche pulse appearing on the sense wire indicates the position of the trajectory.

Important considerations are: whether a uniform field is necessary; what dead-time is allowable, remembering that typically electrons move with a velocity of 5 cm/μs along the

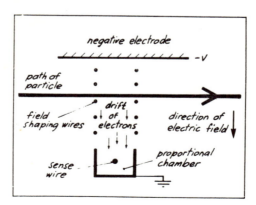

Fig. 10.1.0.1 A basic drift chamber.

drift path; can simultaneous multiple tracks be resolved; how stable do the high voltage, the temperature and the gas composition have to be; what is the performance and cost of the timing electronics, etc. What is clear is that drift chambers are capable of estimating particle-track positions to an accuracy of 0·1 mm, which is superior to that of spark chambers and is likely to be further improved. According to the application, drift paths may be chosen from 1 cm to 50.

The passage of the particle may be registered with fast external scintillation counters which will define zero time. One approach is then to measure the subsequent drift time with a time-to-amplitude converter coupled to an analogue-to-digital converter. Alternatively one may count oscillator pulses with a scaler that is opened by the scintillation pulse and closed by the signal from the anode wire.

10.2 THE CHOICE OF GAS

10.2.1 Drift velocity

The drift velocities for a number of gases are presented as a function of field in Fig. 10.2.1.1. In the drift chamber a linear relationship between time and distance is advantageous and ideally therefore one should aim to use a gas in which the drift velocity is independent of electric field. By this means small irregularities in the drift field will have only a negligible effect on the linearity. In the graph it is seen that flat regions occur for methane, ethylene and some of the mixtures. Velocity saturation at high-field strengths is useful because it minimizes the changes in the velocity of the electrons in the proximity of the sense wires.

The choice of drift velocity is influenced by the length of the drift path, the intensity of particles crossing the chamber, and the speed of the recording electronics. For example, if electrons moving with a velocity of 6 cm/μs have to traverse a 10 cm path, the effective memory time will be 1·7 μs—unless special precautions are taken. A faster velocity would reduce this time but it would also require a shorter electronic resolving time.

The operation of drift chambers requires that electrons move in gases with only small attachment coefficients—otherwise a serious limitation would be imposed on the length of drift path.

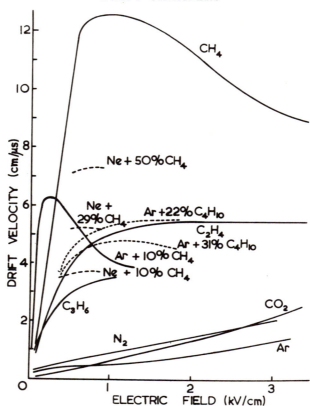

Fig. 10.2.1.1 Electron drift velocities in various gases. (Hurst *et al.*, 1963; Saudinos *et al.*, 1973; Charpak *et al.*, 1973)

The noble gases will not attach, and it appears that ethylene, methane and isobutane are satisfactory. The attenuation in the cloud of drifting electrons may be determined by observing the distribution of pulse heights at the sense wire, as a function of drift path length, or field. Cheng *et al.* (1973) found that the attachment is roughly proportional to the total time spent by the electrons in the gas, e.g. in an argon (13)/Ethylene (87) mixture the pulse height diminished by a factor 2 along a path of 22 cm in a field of 311 V/cm. Saudinos *et al.* (1973) however found no appreciable attenuation in pulse heights over 20 cm with 820 V/cm in a mixture neon (80)/ethylene (15)/isobutane (5).

10.2.2 Electron diffusion

The limit on the spatial resolution attainable with drift chambers is set by the inherent width of the trail of ion pairs, by the diffusion of electrons in flight and by timing inaccuracies. The greatest of these is likely to be the second. The measurement uncertainty will therefore be related to $\sigma = \sqrt{(2Dt)}$ where D is the diffusion coefficient and t the drift time. We have

$$\sigma = \left(2 \int_0^t D(E) \, dt\right)^{\frac{1}{2}}$$
$$= \left(2 \int_0^S \frac{D(E)}{w} \, ds\right)$$
$$= \left(2 \int_0^S \frac{D(E)}{\mu(E)} \frac{ds}{E}\right)^{\frac{1}{2}}$$

where w is the drift velocity and $\mu(E)$ the mobility (see Chapter 1).

To minimize σ one should minimize D/w. Some curves of this ratio against E are displayed for various gases in Fig. 10.2.2.1. (N.B. the actual values should be treated with reserve—they vary considerably in the literature.) As an example, electrons, after drifting in ethylene along a 1 cm path in a field of 1 kV/cm, will have a diffusion width

$$\sigma = \sqrt{2(D/w)l} = \sqrt{2.10^{-4} \cdot .1} \text{ cm} = 0.14 \text{ mm}.$$

Other things being immaterial, one would prefer ethylene to methane because of its smaller diffusion width.

The ideal drift gas has probably not yet been found. One may note the comment (Cheng et al., 1973) that a multi-component gas mixture has D/μ close to the gas with the lower D/μ. The possibility exists of adding special gases. If one refers to Fig. 1.4.2.1 one sees that the addition of a small percentage of nitrogen to argon very significantly increases the drift velocity. In this case the nitrogen effectively cools the electrons—inelastic collisions due to the excitation of vibrational levels in N_2 reduce the average energy of the electrons to about 1 eV (from 10 eV in pure argon at $E/p = 1$). This reduces the average momentum lost per second, which results in an enhancement of the drift velocity and a reduction in D/μ ($=\frac{2}{3}\bar{\varepsilon}$) (Klema and Allen, 1950). At the Frascati conference F. Sauli reported adding 2·5% of

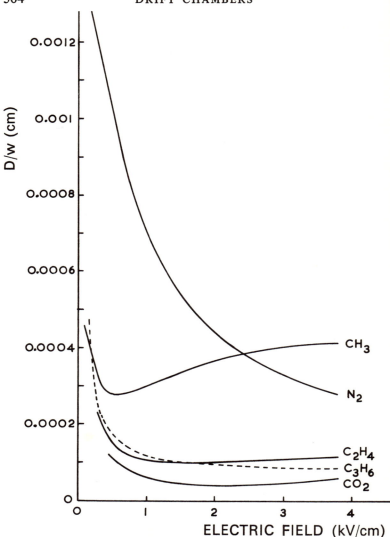

Fig. 10.2.2.1 The ratio of the diffusion coefficient to the drift velocity for various gases at atmospheric pressure as a function of electric field. Values for argon are off the graph; e.g. $D/w = 0.008$ and 0.011 cm for $E = 1000$ and 400 V/cm respectively. The data are taken from Cochran and Forrester (1962) and Warren and Parker (1962).

nitrogen to an argon–isobutane mixture to cool the electrons, and to give a better constant velocity.

In practice many (N) ionization electrons will drift along the path length. If one were to detect the centroid of the group of electrons rather than just the first few electrons to arrive at the sense wire, the precision could be improved by as much as $\sqrt{N}$. This feat, however, would require more sophisticated electronics than the simple threshold discriminations commonly used.

An important point which is often ignored is that electrons diffuse by different amounts in the directions parallel and perpendicular to the electric field (Walenta, 1973). The classical measurements of D/μ refer to the transverse case. However, diffusion parallel to the field (D_L) has been shown by Parker and Lowke (1969) to be much reduced. For example, in argon over the range of E/p from 0·01 to 1, D_T/D_L is about 7; and about 2 in helium, hydrogen and nitrogen. The theory takes account of electron density gradients when the Boltzmann equation is solved. When the electron collision frequency increases with energy, the theory predicts that the leading edge of the pulse has a reduced mobility because of a higher average electron speed and collision rate. Similarly the mobility of the trailing edge is enhanced and consequently the half width of the pulse in the field direction is reduced. D_L/D_T at high fields is found to be approximately 0·5 and 0·2, for momentum transfer cross-sections that are independent of and linearly proportional to the electron energy. When inelastic collisions are included in the theory good agreement with measurements is obtained.

The spatial resolution is related to D/w, which may be written: $D/w = $ const. (η/E), where η is the electron temperature (i.e. the ratio of the agitation energy of free electrons to that of the gas molecules). Hough (1972) has presented a method for calculating the diffusion of electrons in argon/methane mixtures.

10.3 EXAMPLES OF DRIFT CHAMBERS

10.3.1 Development at CERN

In the original realization of a drift chamber by the Charpak group a separate drift region maintaining a uniform field was isolated from a proportional chamber by a wire grid (Bressani *et al.*, 1969). More recent work, however, has concentrated on a design in which the basic structure of the multiwire proportional chamber is kept.

Figure 10.3.1.1(*a*) shows an element of a large chamber in which a central plane consists of anode "sense" wires alternating with "field" wires. The enveloping cathode planes are composed of parallel equidistant wires individually held at potentials chosen to create a well-controlled electric field in the drift space between the sense and field wires. In the figure, equipotentials are shown calculated for a cathode-potential distribution ranging from $V_M = -1$ to $V_m = -0.58$.

A prototype chamber had the following parameters: sense wire diameter and length—20 μm and 30 cm; distance between sense wires—48 mm; distance between cathodes—6 mm; cathode wire diameter and spacing—100 μm and 2 mm. With $V_M = 3.2$ kV, $V_m = 1.5$ kV, the resulting field nowhere dropped

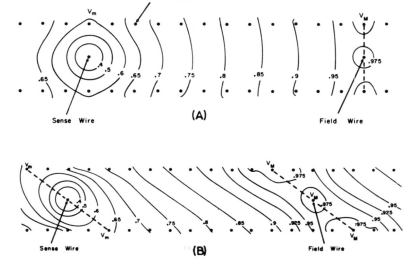

Fig. 10.3.1.1 (A) CERN adjustable field drift chamber. The cathode-wire potentials range from $V_M = -1$ to $V_m = -0.58$, and the sense wire is earthed. The distance between the field and sense wires is 24 mm. (B) Tilting the equipotentials to cope with a magnetic field. (Charpak *et al.*, 1973)

below 600 kV/cm—whereas it did in the absence of a field wire, and also if a uniform-cathode potential was applied. In an argon/isobutane mixture of 75/25, 140 mV pulses were obtained for the Fe^{55} X-ray source, and 30 mV pulses for 1 GeV/c pions. The proportional amplification was satisfactory—yielding a 17% FWHM resolution for the 5·9 keV X-ray. A threshold of detection of −3 mV was found to be sufficient to guarantee full efficiency for relativistic particles.

Fig. 10.3.1.2 Time spectrum in the CERN chamber detecting two narrow beams, 10 and 12 mm from the sense wire. (The scale is about 2 channel/ns.) (Charpak *et al.*, 1973)

To test the space–time relationship, delayed coincidence techniques were applied to a couple of chambers in conjunction with accurately timed scintillation counters. The measured accuracy of track location as a function of path length, after corrections for uncertainties in the zero time and the coincidence widths, was determined. The expected dependence on diffusion was observed. In the argon–isobutane mixture, for a 1 cm path, the accuracy was 120 μm which implied an accuracy of about 400 μm in 10 cm. This corresponds to $D = 0·032$ m^2/sec. Figure 10.3.1.2 shows the time spectrum of pulses from two narrow beams of particles, about 2mm apart, with drift paths of 10 and 12 mm.

10.3.2 The Saclay chamber

Nice chambers with long drift paths have been developed at Saclay (Saudinos *et al.*, 1973). A double chamber with two independent drift paths is shown in Fig. 10.3.2.1. A high tension of −45 kV creates a field of 820 V/cm down the 50 cm drift regions which are terminated by single wire proportional counters. The chambers have a height of 15 cm and have 10 μm thick mylar windows. A high-resistance bridge (8 MΩ/cm) is included within the enclosure. A uniform field is maintained down the

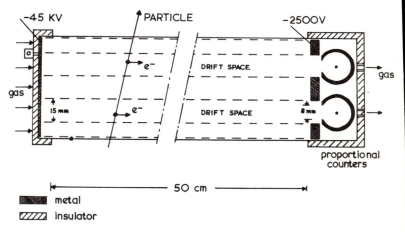

Fig. 10.3.2.1 The Saclay chamber with two drift regions. (Saudinos *et al.*, 1973)

length of the drift path with a set of electrodes held at appropriate potentials between −45 kV and −2·5 kV. These consist of bands of mylar foil, 6 μm thick and aluminized on both sides, which encircle the chamber. The sensing anodes are 40 micron molybdenum wires which are maintained at potentials between 0 and +700 V.

The drift times were measured with a time-to-amplitude converter coupled to a multichannel analyser. Although in early experiments with 25 cm chambers a mixture of 80% neon/15% ethylene/5% isobutane was used, the drift velocity of 4·2 cm/μs was rather slow. With the 50 cm chamber a mixture of 50% neon/50% methane was first used, and subsequently pure methane to take advantage of the faster drift speed (9·7 cm/μs in pure methane).

A study with minimum-ionizing particles resulted in spatial resolutions (FWHM) of 0·4, 0·7, 1·25, 1·3 mm for drift paths of 7, 17, 37, 50 cm respectively. The use of a double-chamber configuration allows one to reject fortuitous-angled trajectories with gating techniques. The distribution of ionization losses of 1·04 GeV protons in the sensitive regions was obtained by measuring the distribution of pulse heights from the sensing wire—and it was found that the distributions differed imperceptibly for drift paths of 5 cm and 25 cm, suggesting no attenuation in the number of drifting electrons over a distance of 20 cm!

10.3.3 The Heidelberg chamber

A most important step in the development of drift chambers was the multiwire chamber designed at Heidelberg (Walenta *et al.*, 1971). In Fig. 10.3.3.1, the wires Y are anodes (earthed) and the wires X are cathodes (-2200 V). Electrons from ionization sites on a trajectory intersecting line XY, length 10 mm, will

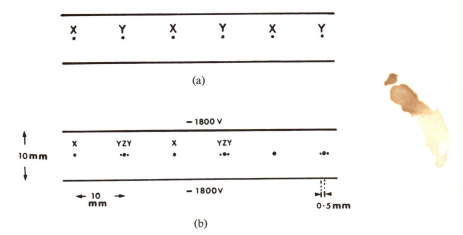

Fig. 10.3.3.1 Multiwire geometries at Heidelberg. Wires X are field wires, wires Y are sensing anodes and a uniform potential is maintained on the enveloping cathodes. (Walenta *et al.*, 1971)

drift to the anode, and the delay before avalanching will indicate their origin.

If only alternate wires X and Y were present, a pulse on any anode Y would not distinguish trajectories to the left and to the right. To overcome this ambiguity, two closely spaced anode wires (Y) were used, with an intermediate shielding wire (Z) at -2200 V.

Chambers have been operated with dimensions 80×130 cm^2 and 60×200 cm^2. With 86% argon, 10% methane and 4% isobutane, the maximum drift time for 10 mm was 230 ns and an accuracy of $\sigma = 0\cdot3$ mm was achieved.

Later, Walenta (1973) reported using the mixture 84% argon/ 9% methane/7% isobutane, which gave a nearly linear time–distance relationship. With this geometry of two closely placed sensing wires, quenching is very important, otherwise an avalanche on one wire will propagate to the other. In general it is advantageous to decrease the isobutane to allow lower operating voltages, but 7% was the lowest satisfactory limit. Continuous mixing of the gases with flow meters was found to be best.

10.3.4 The giant Harvard chambers

Magnificent 4 m × 4 m drift chambers have been built at Harvard (Cheng *et al.*, 1973). With only thirty-four sensing wires a spatial resolution of one part in 10,000 (i.e. 0·4 mm in 4 m) is achieved.

A simple modular configuration was chosen. The metallic 4×4 m^2 cathode walls are separated by 5 cm—a depth partly determined by the necessary size of the supporting frame. The central plane is composed of alternate sensing and field wires. Effectively a number of drift cells are created, each consisting of a sense wire with a 5 cm drift region to either side, and sharply limited by the field wires. And each cell spans the 4 m.

Gold-plated 100 μm molybdenum field and sense wires were used. Such wires sag <0·2 mm over 4 m when placed horizontally under a tension of 150 gm. The onset of electrostatic instabilities suggests (see Chapter 7 and Cheng *et al.*, 1973) that the maximum allowable wire length is

$$L_{max} = \frac{4d \ln(d/r_0)}{V} \sqrt{T}$$

where d is the distance from the wire to the earth plane (cm), r_0, V, T the wire radius (cm), potential (statvolts) and tension (dynes). For example, with $V = 6$ kV, $r_0 = 50$ μm, $d = 25$ mm, $T = 150$ gm, $L_{max} = 12$ m—well above the 4 m of the Harvard chambers.

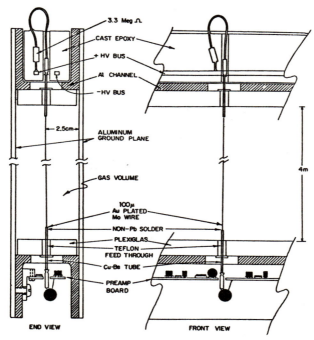

Fig. 10.3.4.1 Constructional details of the giant Havard drift chambers. (Cheng *et al.*, 1973)

Some details of the construction are shown in Fig. 10.3.4.1. The frame consists of square-welded U-profile aluminium channel reinforced by two I-beams crossing the chamber at 1/3 and 2/3 positions. Aluminium sheets are epoxied to the frames to enclose the gas, etc. A plexiglass insulator is epoxied to the interior of the frame and the wires pass through Teflon feedthroughs via thin tubes to suppress corona. As a precaution, all wires are tested for corona discharge to 10 kV before closing the chamber.

Ethylene was chosen to be the basic operating gas because it has a good D/μ ratio and because the electron velocity saturates at high fields. Some proportion of argon (typically 20%) was added to reduce the working voltage of the chamber and stabilize the gas. Typical voltages were +6 kV on the sense wires, −1 to −2·5 kV on the field wires, and the outer walls earthed. The pulse rise and fall times were 50 and 700 ns—which implies a double-pulse resolution of 200 ns if differentiation is employed.

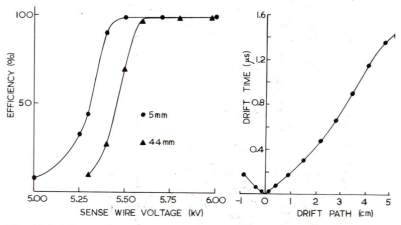

Fig. 10.3.4.2 Performance of the large Harvard drift chambers filled with argon (20%)/ethylene (80%). (*Left*) High-voltage efficiency curves for a proton beam incident at 5 mm and 44 mm from the sense wire. (*Right*) Drift time as a function of proton beam distance from the sense wire, with +6 kV on the sense wire and −1 kV on the drift wire. (Cheng *et al.*, 1973)

The simple geometry means a varying field over the drift region and consequently a varying drift velocity. One cannot rely on a strictly linear relationship and it is necessary to calibrate measured drift times against trajectory positions. In 80% C_2H_4/20% argon, 100% efficiencies were obtained for sense wire voltages above 5·8 kV, and the plateau was 1·5 kV long (Fig. 10.3.4.2).

Good Landau distributions were observed regardless of the beam position, and the overall spatial resolution was 0·35 mm.

10.4 READ-OUT ELECTRONICS

10.4.1 Introduction

Drift chambers require very fast electronics if their optimum performance is to be approached. As an example, consider electrons drifting at a velocity of 4 cm/μs over a path length of 1 cm. The diffusion width in this case is about 100 μm which is equivalent to a time spread of 2·5 ns. Drift times should therefore be recorded to an accuracy better than 2·5 ns. This calls for counting frequencies of about 500 MHz which is much faster than that required for proportional chambers. The faster electronics will be expensive, but on the other hand, there are fewer wires to handle.

10.4.2 The digital approach

This method relies on recording a number of standard oscillator pulses in a scaler that is open for a period related to the drift time. The Heidelberg group successfully employed this technique (Schürlein et al., 1973).

One possible scheme is that proposed by Dr. H. Verweij of CERN (1973). Although the signalling wire is identified, for the purposes of time measurement the sense wires are economically grouped together in eights (Fig. 10.4.2.1). The actual number of wires grouped together would depend on the particle intensities.

The wire signals are amplified by a preamplifier mounted on the chamber and sent via a twisted pair or coaxial cable to the "line receiver, trigger and derandomiser". There the signal is standardized and synchronized to the clock. A "busy" signal activates the system. A "start" signal from the scintillation counter makes all scalers start counting the clock. When one of the wires fires, the processed signal is fed to a general stop OR-gate which drives the stop line of the scalers. In parallel, this signal is also connected to the wire number OR-gates. The first stop that arrives stops the first scaler and sets the "wire number" flip/flop in accordance with the wire that produced the stop. Simultaneously it will *enable* the second channel (scaler and wire number input gates), which is then ready for the second stop, etc. Immediately after a scaler has been stopped the data can be read out.

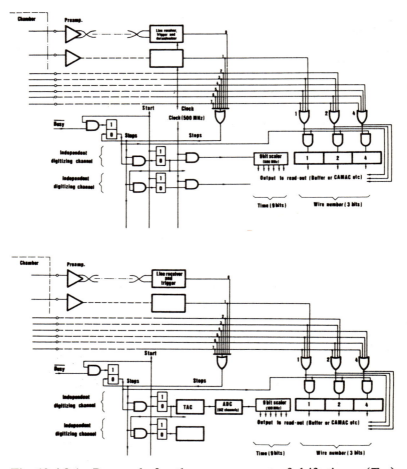

Fig. 10.4.2.1 Proposals for the measurement of drift times. (*Top*) a digital scheme; (*bottom*) an analogue conversion scheme. (Verweij, 1973)

With a clock frequency of 500 MHz, a time resolution of ± 2 ns is obtainable. To cover drift times up to 1024 ns (2×512) a nine-bit scaler is needed.

10.4.3 The analogue method

In this approach the drift times are measured by converting the time duration to a proportionate voltage amplitude. This is done with a unit called a time-to-amplitude converter (TAC). The voltage is then digitized with an analogue-to-digital converter (ADC). A typical ADC with 256 channels gives a 2·4 ns resolution for a full range of 625 ns. The Saclay group (Saudinos *et al.*, 1973), has used this analogue method.

In Fig. 10.4.2.1(*b*) the digital proposal has been modified by replacing the scalers with a TAC, an ADC and a scaler. The way of activating the independent channels is the same as the digital case. The processing circuit can be simpler because there is no need to synchronize the wire pulse with a clock. A disadvantage of this approach is that after a stop has occurred the data are not immediately available because the analogue signal must be converted. The conversion time for a run-down type of ADC with, say, a 100 MHz clock and 512 channels will be a maximum of 5·12 μs. An advantage is that the time resolution can be increased by increasing the number of ADC channels—if the TAC and ADC are stable enough (Verweij, 1973).

10.4.4 Time stretching

The digital measurement of drift time intervals with 500 MHz circuits can be expensive. A simpler technique is first to multiply the time intervals by a constant factor, and then to measure with relatively slow circuits. This multiplication can be achieved by discharging a charged capacitor with a high current (i_2) for the real time interval and subsequently allowing it to charge again at a low current (i_1). The charge time will be (i_2/i_1), times the drift time, which can easily be measured if, say, the ratio (i_2/i_1) is 20. This technique has been used in the Harvard instrumentation (Cheng *et al.*, 1973).

Figure 10.4.4.1 illustrates the operation of a time-stretcher. In the quiescent state, transistor $T1$ is on, $T2$ is off, $T3$ is on, the voltage at A is $V_0 + 0.7$ V, and the output is high. At t_1 the input falls to 0 V, $T1$ goes off and $T2$ comes on. The capacitor is drained by a current of ($\alpha - 1$)i and A drops linearly in voltage. $T3$ goes off, and the output goes to 0V. At t_2 the input returns high, $T1$ comes on and $T2$ goes off. The capacitor is charged by the current i and A slowly rises linearly in voltage. At t_3 A has returned to $V_0 + 0.7$ V, $T3$ comes on and the output returns to $+3$ V (Cheng et al., 1973). In the Harvard case, alpha was chosen to be 6, and standard 20 MHz TTL circuitry was employed in the digitizer stage.

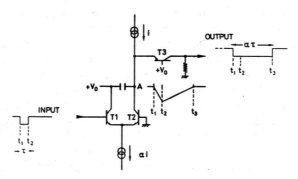

Fig. 10.4.4.1 Simplified diagram of a time stretcher. (Cheng et al., 1973)

Rubbia (1973), in a practical proposal, suggests that the time-stretching should yield resolutions equivalent to 1000 MHz— for a component cost of 20 Swiss francs.

10.4.5 Reading the second coordinate

The drift time before the avalanche occurs at the sense wire gives one coordinate of the particle trajectory. It would be a great benefit if the second coordinate could be determined simultaneously by measuring the position of the avalanche along the sense wire. Even if a high precision cannot be obtained, a crude location might be well worthwhile; ambiguities created by multiple tracks might be eliminated. A measurement is certainly feasible: it is known that an avalanche on a wire is normally well localized—to within perhaps 0.1 mm (Borkowski and Kopp, 1970).

One approach, the current division method, was proposed ten years ago by Charpak (see section 4.7). If the sense wire is resistive, and if the ends are earthed, then the discharge current that flows from the avalanche to the two ends will be divided according to the path lengths. Referring to Fig. 4.7.0.1, if the length of wire is 1, and x the avalanche distance from one end, we have

$$x = \frac{I_2}{I_1 + I_2} \cdot 1,$$

and

$$1 - x = \frac{I_1}{I_1 + I_2} \cdot 1.$$

Unlike the spark-chamber case, with single avalanches one cannot rely on $I_1 + I_2 =$ constant. Rather, the amount of charge developed varies widely from avalanche to avalanche and so it is sensible to try to measure the ratio $I_2/(I_1 + I_2)$. In practice this is most easily done by taking the ratio of the currents flowing at each end.

Working with a 30 cm long single-wire proportional counter, Kuhlman et al. (1966) showed that avalanches could be located to within 1·2 mm. Hough (1972) has analysed the limits that can be expected with this type of measurement.

A major difficulty is that, for good sensitivity, the impedance of the wire should be much greater than that of the current measuring devices. For fast large-scale detectors, therefore, effort has been concentrated on using normal wires. Charpak et al. (1973) have tried 20 μm tungsten wires (0·25 Ω/cm), and using low-impedance (2 Ω) amplifiers and a fast divider circuit they obtained an accuracy σ of 4 mm. The time to determine the ratio V_1/V_2 was 5 μs. With 50 μm constantan (3 Ω/cm) an accuracy of 2·5 mm in 20 cm was achieved.

Foeth et al. (1973) have studied the currents in an 11 cm long 10 μm tungsten sense wire (7·2 Ω/cm). One end of the sense wire was connected to a low-input impedance (4 Ω) amplifier composed of a ground-base transistor followed by a conventional monolithic wide-band amplifier chip (μA733 by Fairchild). Apparently, without taking the ratio of currents, but rather just observing the amplitude at one end, they found they could measure to a precision of about 6 mm FWHM.

Another possibility which at present shows promise is the use of delay lines (Charpak, 1973). The principle has been successfully applied to proportional chambers (Grove *et al.*, 1973) (see section 7.6.2). Charpak and Sauli have explored the incorporation of the delay line in the cathode of a chamber. In their tests, they replaced three cathode wires of a multiwire chamber with a 4 mm wide plastic foil, coated on one side with a 10 μm layer of copper and on the other side with a zigzag copper line of 9 lines/mm. The line has the characteristics of 1 cm/ns, 65 Ω impedance and 4 ns rise-time. The avalanche induces a signal which propagates to both ends of the line and the time difference is measured. With a 20 cm line, the avalanche position along the cathode could be ascertained to $\sigma = 2$ mm.

Another proposal by the CERN group is to have, facing the sense wires, rather broad strips on which induced-pulse heights may be measured. The centre of gravity of the pulses induced on the strips will indicate the avalanche coordinate. The strips, say 4 mm wide, may be placed parallel or orthogonal to the sense wires, and accuracies of 100 μm are possible.

A proposal (Rice-Evans, 1973) that remains to be explored is a chamber in which the drift time determines one coordinate of a trajectory, and the firing wire in a multiwire array the other. Such a design could be achieved with a large-area drift plane terminated along one edge by a slit behind which lies a multiwire counter with its wires orthogonal to the drift plane.

10.5 PERFORMANCE

10.5.1 Accuracy

The precision obtained with a drift chamber is very much a function of the geometry; long drift paths inevitably mean increased uncertainties due to thermal diffusion. Inaccuracies arise in the timing measurements due to uncertainties in the zero time definition by the scintillation counters (say), the resolving times of the coincidence circuits, and the jitter if simple threshold discriminators are used to detect the sense-wire pulses. Multiple scattering in the chamber windows (say 12 μm, mylar) and in the wires can also contribute.

Other potential causes of uncertainty are fluctuations in the deposition of primary ionization; departures from a linear drift-time/drift-length relation; and geometrical uncertainties such as the exact positioning of the wires.

Walenta (1973) estimates a location accuracy of $\sigma = 0\cdot 2$ mm for the Heidelberg chambers. Charpak *et al.* (1973) obtained 40 μm for tracks close to the sense wires, and 180 μm for drift paths of 24 mm.

10.5.2 Angled tracks and left–right ambiguities

In some configurations—e.g. the multiwire format of Fig. 10.3.3.1(A)—it is impossible to distinguish between particles passing to the left and to the right of a sensing wire. We have seen that the solution adopted at Heidelberg is to employ two closely placed sensing wires with an intermediate potential wire to reduce the electrostatic instabilities. This method however leads to some difficulties when tracks pass between any pair of sense wires.

The approach suggested by Charpak *et al.* (1973) is to employ a stack of three chambers. In Fig. 10.5.2.1, three chambers all with their wires parallel are so disposed that the central chamber has its wires displaced by half a wire spacing with respect to the other two. Clearly no left–right ambiguities arise with three signals appearing.

With this configuration, for a particle traversing at time t_0

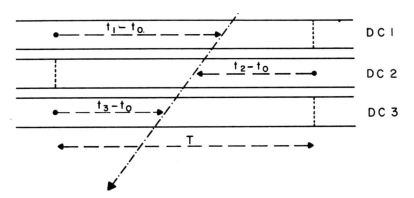

Fig. 10.5.2.1 Removal of left–right ambiguities with a stack of three multiwire drift chambers. The central chamber is displaced so that its sense wires are aligned with the field wires in the other chambers. (Charpak *et al.*, 1973)

and for the indicated drift times, the following relation holds:

$$t_0 - t_2 = \frac{(t_1 - t_2) + (t_3 - t_2)}{4} - \frac{T}{2}$$

where T is the maximum drift time. It is thus sensible to use the central chamber for drift-time definition, and it will be given in terms of $t_1 - t_2$ and $t_3 - t_2$, i.e. eliminating the need for identifying zero time t_0.

A variety of combinations of chambers may be envisaged. One, considered by the Harvard group, is to use three multiwire chambers, each oriented at 60° with respect to the others. This is good for removing multiparticle ambiguities. In fact, the group favoured super-modules, each composed of two pairs of planes, each pair having its wires displaced by half a cell (as in Fig. 10.5.2.1). A knowledge of the four absolute drift times then allows the calculation of the angular directions of the trajectories. Successive super-modules, tilted with respect to each other, must be employed if several particles will traverse the system simultaneously (Cheng et al., 1973).

10.5.3 Time resolution and multiple tracks

Chambers with long drift paths may be plagued with spurious signals. For example, in the Saclay chamber, the maximum drift time is about 7 μs and any noise or parasitic signal during this time will confuse a real event. However, by employing several double chambers in a stack, and imposing a coincidence ($\tau = 60$ ns) condition on the signals, the chance coincidence rate was reduced to negligible proportions even with parasitic rates in excess of 5.10^5/s (Saudinos et al., 1973).

A multiwire format is more appropriate for high intensities of particles. It is advantageous then to treat each drift region independently and in this manner many simultaneous particles can be easily handled.

On the question of what is the minimum detectable separation of two tracks that pass through one drift region, two factors are crucial. One is the electronic resolution and the other is the dead-time of the sensing wire as it recovers from the first avalanche. The electronic limit may be said to be the width of the pulse delivered from the wire amplifier to the following discriminator. Charpak et al. (1973), using a low-impedance (50 Ω) amplifier, got an occupation time not exceeding 50 ns; which implies that two tracks could be resolved if their separation were 2 mm or more—in a direction perpendicular to the wire.

The dead-time due to avalanching is related to the time to collect all the electrons liberated (along a trajectory) in the drift space. In a typical multiwire proportional chamber this amounts to about 200 ns, and the dead-time will be of this order. There is, however, another dead-time which is limited to the actual site of the avalanche; Makowski and Sadoulet (1973), using a high-gain magic gas, have observed dead-times as long as 10 μs on spots localized to within 0·2–0·3 mm on the wire. In drift chambers the effective resolving times remain to be measured.

10.5.4 Use in magnetic fields

Frequently, detection chambers have to be placed in magnetic fields. It is obvious that in a drift chamber, any component of a magnetic field perpendicular to the drift line will affect the motion of the electrons. The effect will be worst when the field is parallel to the wires, in which case the electrons will be deviated to the sides of the chamber and lost.

Townsend (1912) has shown that the lateral displacement of electrons drifting in a gas in an electric field E for a time t is given by:

$$\Delta x = \frac{E}{B} \frac{v^2 \tau^2}{1 + v^2 \tau^2} t$$

where $v = eB/m$ is the Larmor frequency, B the magnetic induction perpendicular to E, and τ the mean free time between collisions for electrons.

The problem has been examined by Charpak et al. (1973). In a drift chamber, the displacement will be greater for longer drift times; and indeed it is found that the efficiency of a multiwire chamber in a magnetic field declines as the trajectory is moved away from the sense wire. One way to diminish the displacement is to reduce τ, i.e. to reduce the mean free path, which implies increasing the gas density.

However, in strong magnetic fields a more radical solution is required. Charpak et al. have experimented with the arrangement shown in Fig. 10.3.3.1(b). By using separate voltage distributions on the two multiwire cathodes, the electric field may be tilted to provide a force to compensate for the Lorentz displacement. If E be the electric field in the direction of the cathode planes, and B the magnetic-induction perpendicular to E, the tilting angle necessary to oblige the electrons to drift parallel to the cathode planes is

$$\alpha = \arctan \frac{Bw}{E}.$$

With $E = 600$ V/cm, $B = 15$ kG and $w = 4.10^4$ m/s, $\alpha = 45°$. Charpak and his group achieved this tilt by connecting the two cathode planes to independent power supplies. With optimum voltages, a 100% efficiency was achieved over the whole drift length.

APPENDIX A

High-voltage Properties of Materials

THE design of high-voltage equipment is governed largely by the materials to be employed. The properties of these materials are not well documented and in any case they are likely to vary with the circumstances. Nevertheless Bulos *et al.* (1967) have summarized some of the data—for example, the breakdown fields for various gases and liquids under d.c. and pulsed conditions. With solids they found evidence of fatigue: whereas lucite could withstand 12 MV/in. it could also be punctured with a series of 200 ns pulses at 400 kV/in.! In Fig. A1 some results of high-voltage single-shot impulse breakdown through thin sheets are presented, and Bulos *et al.* suggest they are equivalent to d.c. breakdown conditions.

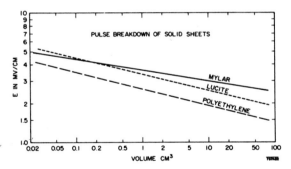

Fig. A1 Electric field for pulse breakdown in thin sheets versus sample volume (unpublished data of J. C. Martin and I. Smith, reported by Bulos *et al.*, 1967)

APPENDIX B

Preparation of Plastic Scintillators

SPARK chambers are often triggered by signals from fast plastic-scintillation counters. In many cases, large sheets of scintillator are necessary—with dimensions of the order of metres—and good performance demands a close attention to detail in their construction. The craft of preparation of these counters has been finely developed at CERN by M. Leslie Thornhill and his colleagues, and in the following we shall outline a successful procedure.

Typically, the scintillator is coupled to the photomultiplier with a perspex (lucite) light guide. This may be achieved with one triangular piece or with many separate strip guides. In the latter case, in which the strips are rotated out of the scintillator plane and brought together at the photomultiplier, the light output may be improved by a factor of 3 to 5.

The perspex must be a good grade (ICI, optical or aircraft quality, with ultra-violet transmittance) and it is most important that it is given the highest attainable polish (paste from Plexbril, Industriestrasse, 8108 Dällikon, Zürich). The scintillator (Nuclear Enterprises, Sighthill, Edinburgh) may be polished with this paste too, using a soft cloth wheel (max. 1500 r.p.m.) and a very light touch to avoid the heating and crazing of the plastic. The sticking of plastic to perspex is done with Eastman glue 910 which may harden in a few minutes and does not yellow due to radiation damage in an intense particle beam (cf. epoxy resin). This glue allows a later clean separation of perspex and plastic, if necessary. The plastic surface to be glued should not be polished but rendered flat with 3 M Wetordry tri-m-ite paper (A wt Type W, waterproof silicon carbide). Gluing of perspex to perspex is best done with ICI tensol cement No. 7.

The plastic and light guide should be sheathed in aluminium foil to help reflect the light. Well-rolled, pure aluminium is best. Finally, the exterior is covered with black tape.

Bibliography

Albrow, M. G., private communication (1970).
Albrow, M. G., Barber, D. P., Bogaerts, A., Bošnjaković, B., Brooks, J. R., Clegg, A. B., Erné, F. C., Gee, C., Kanaris, A. D., Lacourt, A., Locke, D. H., Murphy, P. G., Rudge, A., Sens, J. C., and van der Veen, F., *Phys. Lett.* **40B**, 136 (1972).
Alikhanyan, A. I., Asatiani, T. L., and Materoyan, E. M., *Zhur. Eksp. i. Teoret. Fiz.* **44**, 773 (1963) [*Sov. Phys. J.E.T.P.* **17**, 522 (1963)].
Alikhanian, A. I., Avakina, K., Garibian, G., Lorikian, M., and Shikhliarov, K., *Phys. Rev. Letters*, **25**, 635 (1970).
Alikhanyan, A. I., Asatiani T. L., Avakyan, K., Zhirova, L., Ivanov, V., Krishchyan, V., Matevosyan, E., *Sov. Phys. J.E.T.P.* **34**, 677 (1972).
Alikhanov, A. I., Lubimov, V., and Eliseev, G. P., *Proc. Int. Conf. on Instrumentation for High-Energy Physics*, CERN, p. 87 (1956).
Alleyn, H., Guezennec, J. J., and Muratori, G., CERN Report, CERN 68-34 (1968).
Allkofer, O. C., Bagge, E., Henning, P. G., and Schmieder, L., *Atomkernergie*, **2**, 88 (1957).
Allkofer, O. C., with Dau, W. D., and Grupen, C., *Spark Chambers* (Verlag Karl Thiemig KG., Munchen, 1969).
Allkofer, O. C., Grupen, C., and Maxion, G., *Nuc. Inst. Meth.* **79**, 181 (1970).
Alvarez, L., Anderson, J., El Bedwei, F., Burkhard, J., Fakhry, A., Girgis, A., Goneid, A., Hassan, F., Iverson, D., Lynch, G., Miligy, Z., Moussa, A. H., Mohammed-Sharkawi, Yazolino, L., *Science*, **167**, 832 (1970).
Amato, G., and Petrucci, G., CERN Report, CERN 68-33 (1968).
Anderson, H. L., CERN Report, CERN 64-30, p. 92 (1964).
Anderson, L., Radermacher, E., and Rubbia, C., *Nuc. Inst. Meth.* **75**, 341 (1969).
Aronson, S., Catura, R., Chen, H., and Chen, K., *Nuc. Inst. Meth.* **46**, 93 (1967).
Asatiani, T. L., Gazaryan, K. A., Ivanov, V. A., Zhmyrov, V. N., and Nazaryan, A. A., *JETP Letters*, **6**, 83 (1967) and **6**, 171 (1967).
Astbury, P., Finocchiaro, G., Michelini, A., Websdale, D., West, C.H., Beusch, W., Gobbi, B., Pepin, M., Polgar, E., and Pouchon, M. A., *Nuc. Inst. Meth.* **46**, 61 (1967).

Autonès, P., Bareyre, P., Gaillard, J. M., Robert, G., and Seige, R., *Nuc. Inst. Meth.* **24**, 418 (1963).
Bacon, F. F., and Nash, W. F., *Nuc. Inst. Meth.* **37**, 43 (1965).
Badareu, E., and Popescu, I., *Gaz Ionisés* (Dunod, Paris, 1965).
Bainbridge, G. R., and Prowse, W. A., *Can. J. Phys.* **34**, 1038 (1965).
Banaigs, J., Duflo, J., Goldzahl, L., Mouellic, B., Delorme, C., Kiessler, R., and Michaelis, E. G., *Nuc. Inst. Meth.* **26**, 137 (1964).
Bardin, R. K., Gollon, P. J., Ullman, J. P., and Wu, C. S., *Phys. Lett.* **26B**, 112 (1967).
Bardon, M., Lee, J., Norton, P., Peoples, J., and Sachs, A. M., CERN Report, CERN 64–30, p. 41 (1964).
Bates, D. R. (ed.), *Atomic and Molecular Processes* (Academic Press, New York, 1962).
Bayle, P., and Schmied, H., CERN Report, CERN 72–9 (1972).
Beall, E. F., Cork, B., Murphy, P. G., and Wenzel, W. A., *Nuc. Inst. Meth.* **20**, 205 (1963).
Bella, F., Franzinetti, C., and Lee, D. W., *Nuovo Cimento*, **10**, 1338 (1953) and **10**, 1461 (1953).
Bemporad, C., Beusch, W., Melissinos, A. C., Schuler, E., Astbury, P., and Lee, J. G., *Nuc. Inst. Meth.* **80**, 205 (1970).
Benz, H., Chikovani, G., Damgaard, G., Focacci, M., Kienzle, W., Lechanoine, C., Martin, M., Nef, C., Schübelin, P., Baud, R., Bošnjaković, B., Cotteron, J., Klanner, R., and Weitsch, A., *Phys. Lett.* **28B**, 233 (1968).
Bevan, A. R., *Nature*, **164**, 454 (1949).
Binon, F., Bobyr, V., Duteil, P., Gouanere, M., Hugon, L., Spighel, M., and Stroot, J., *Nuc. Inst. Meth.* **94**, 27 (1971).
Blüm, P., and Brinckmann, P., *Nuc. Inst. Meth.* **75**, 335 (1969).
Blunck, O., and Liesegang, S., *Z. Physik*, **128**, 500 (1950).
Blunck, O., and Westphal, K., *Z. Physik*, **130**, 641 (1951).
Board, S. J., Dean, A. J., and Ramsden, D., *Nuc. Inst. Meth.* **65**, 141 (1968).
Bogdanova, I. P., Bochkova, O. P., and Frish, S. E., *Sov. Phys. Doklady*, **9**, 370 (1964).
Bohmer, V., and Schopper, H., CERN Courier, **11**, 231 (1971).
Bolotov, V. N., and Devishev, M. I., *Nuc. Inst. Meth.* **31**, 213 (1964).
Bolotov, V. N., Devishev, M. I., and Klimanova, L. F., *Nuc. Inst. Meth.* **44**, 77 (1966).
Bondarenko, G. B., and Dolgoshein, B. A., *Coll. Int. Electronique Nucléaire* (Versailles, 1968).
Borkowski, C. J., and Kopp, M. K., *I.E.E.E. Trans. Nuc. Sci.* NS–19, No. 3, 161 (1972).
Bortner, T. E., Hurst, G. S., and Stone, W. G., *Rev. Sci. Inst.* **28**, 103 (1957).

Bouclier, R., Charpak, G., Dimčovski, Z., and Sauli, F., CERN N.P. Internal Report 70-2 (1970).
Bouclier, R., Charpak, G., Dimčovski, Z., Fischer, G., Sauli, F., Coignet, G., and Flügge, G., *Nuc. Inst. Meth.* (submitted June 1970).
Bowe, J. C., *Phys. Rev.* **117**, 1411 (1960).
Bradamante, F., and Sauli, F., *Nuc. Inst. Meth.* **56**, 268 (1967).
Bradbury, N. E., and Tatel, H. E., *J. Chem. Phys.* **2**, 827 and 835 (1934).
Branscomb, L. M., *Atomic and Molecular Processes* (ed. Bates, D. R.) (Academic Press, New York, 1962).
Bressani, T., Charpak, G., Favier, J., Massonnet, L., Meyerhof, W. E., and Zupančič, C., *Nuc. Inst. Meth.* **68**, 13 (1969).
Bressani, T., Charpak, G., Rahm, D., and Zupančič, C., Dubna meeting on filmless and streamer chambers (1969).
Brinckmann, P., *Nuc. Inst. Meth.* **67**, 352 (1969).
Brown, J. L., Glaser, D. A., and Perl, M. L., *Phys. Rev.* **102**, 586 (1956).
Brown, S. C., *Basic Data of Plasma Physics* (Wiley, New York, 1959).
Browne, P. F., *Proc. Phys. Soc.* **86**, 1323 (1965).
Bruining, H., *Physics and Applications of Secondary Electron Emission* (Pergamon, London, 1954).
Budini, P., Taffara, L., and Viola, C., *Nuovo Cimento*, **18**, 864 (1960).
Buhler-Broglin, A., Fortunato, G., Massam, T., Muller, T., and Zichichi, A., *Nuovo Cimento*, **45**, 520 (1966).
Bulos, F., Odian, A., Villa, F., and Yount, D., SLAC Report No. 74, 1967 (Stanford).
Burleson, G. R., Hoang, T. F., Kalmus, P. I. P., Kuskowski, R. L., Niemela, L. Q., Roberts, A., Romanowski, T. A., Warshaw, S. D., and Yurka, G. E., *Nuc. Inst. Meth.* **20**, 185 (1963).
Burnham, J. U., Rogers, I. W., Thompson, M. G., and Wolfendale, A. W., *J. Sci. Inst.* **40**, 296 (1963).
Burnham, J. U., and Thompson, M. G., *J. Sci. Inst.* **41**, 108 (1964).
Byrne, J., Shaikh, F., and Kyles, J., *Nuc. Inst. Meth.* **79**, 286 (1970).
Byrne, J., *Nuc. Inst. Meth.* **74**, 291 (1969).
Caris, L., Kuiper, B., and Williams, E. M., *Nuc. Inst. Meth.* **59**, 145 (1968).
Catura, R., and Chen, K., *Nuc. Inst. Meth.* **36**, 317 (1965).
Cavelleri, C., Gatti, E., and Redaelli, C., *Nuc. Inst. Meth.* **20**, 238 (1963).
Chaminade, R., Da Chazeaubeneix, J. C., Fontaine, J. M., Garreta, D., Laspalles, C., Saudinos, J., and Van den Bossche, M., CEN Saclag Unpublished Report (quoted by Charpak, 1971).
Chang, W. Y., and Rosenblum, S., *Phys. Rev.* **67**, 222 (1945).

Charpak, G., *J. Phys. Radium*, **18**, 539.
Charpak, G., *Nuc. Inst. Meth.* **15**, 318 (1962).
Charpak, G., Favier, J., and Massonnet, L., *Nuc. Inst. Meth.* **24**, 501 (1963).
Charpak, G., Massonnet, L., and Favier, J., *Progr. in Nuclear Techniques and Instrumentation*, **1**, 323 (1965).
Charpak, G., Favier, J., and Massonnet, L., *Proc. Int. Conf. on Instrumentation for High-Energy Physics*, Stanford, p. 3 (1966).
Charpak, G., Bouclier, R., Bressani, T., Favier, J., and Zupančič, C., *Nuc. Inst. Meth.* **62**, 262 (1968) and **65**, 217 (1968).
Charpak, G., *Ann. Rev. Nuc. Sci.* **20** (1970).
Charpak, G., Rahm, D., and Steiner, H., *Nuc. Inst. Meth.* **80**, 13 (1970).
Charpak, G., and Sauli, F., *Nuc. Inst. Meth.* **96**, 363 (1971).
Charpak, G., Breskin, A., and Piuz, F., *Nuc. Inst. Meth.* **100**, 157 (1972).
Charpak, G., *I.E.E.E. Trans. Nuc. Sci.* **NS-19**, No. 3, 152 (1973).
Charpak, G., International Conference, Frascati (1973).
Charpak, G., Sauli, F., and Duinker, W., *Nuc. Inst. Meth.* (1973) (to be published).
Cheng, D. C., Kozanecki, W. A., Piccioni, R. L., Rubbia, C., Sulak, L. R., Weedon, H. J., and Whittaker, J., *Nuc. Inst. Meth.* (1973).
Chikovani, G. E., Mikhailov, V. A., and Roinishvili, V. N., *Phys. Letters*, **6**, 254 (1963).
Chikovani, G. E., Roinishvili, V. N., and Mikhailov, V. A., *Nuc. Inst. Meth.* **29**, 261 (1964).
Chikovani, G. E., Roinishvili, V. N., Mikhailov, V. A., and Javrishvili, A. K., *Nuc. Inst. Meth.* **35**, 197 (1965).
Chikovani, G. E., Laverrière, G. C., and Schübelin, P., *Nuc. Inst. Meth.* **47**, 273 (1967).
Christenson, J. H., Clark, A. R., and Cronin, J. W., *I.E.E.E. Trans. Nuc. Sci.* **NS-11**, No. 3, 310 (1964).
Cohen, M. H., and Lekner, J., *Phys. Rev.* **158**, 305 (1967).
Cochran, L. W., and Forrester, D. W., *Phys. Rev.* **126**, 1786 (1962).
Connor, R. D., *Proc. Phys. Soc.* **B64**, 30 (1951).
Conversi, M., and Gozzini, A., *Nuovo Cimento*, **2**, 189 (1955).
Conversi, M., Focardi, S., Franzinetti, C., Gozzini, A., and Murtas, P., *Nuovo Cimento*, **4**, 234 (1956).
Conversi, M., Giannoli, G., and Spillantini, P., *Lett. Nuovo Cimento*, **3**, 483 (1972).
Conversi, M., *Nat. Phys. Sci.* **24**, 160 (1973).
Coombes, R., Fryberger, D., Piccioni, R., Porat, D., and Dorfan, D., *I.E.E.E. Trans. Nuc. Sci.* **NS-17**, No. 3, 50 (1970).
C.P.S. *User's Handbook*, I (CERN, 1969).

BIBLIOGRAPHY 389

Corrigan, S. J. B., and von Engel, A., *Proc. Roy. Soc.* **A245**, 335 (1958).
Coxell, H., and Wolfendale, A. W., *Proc. Phys. Soc.* **75**, 378 (1960).
Craggs, J. D., and Meek, J. M., *High-Voltage Laboratory Technique* (Butterworths, London, 1954).
Cranshaw, T. E., and de Beer, J. F., *Nuovo Cimento*, **5**, 1107 (1957).
Crawford, J. F., Dobinson, R. W., Osmon, P. E., and Strong, J. A., *Nuc. Inst. Meth.* **52**, 213 (1967).
Cronin, J. W., and Renniger, G., *Rev. Sci. Inst.* **32**, 496 (1961).
Cronin, J. W., and Renniger, G., *Proc. Int. Conf. on Instrumentation for High-Energy Physics*, p. 271 (Berkeley, 1960; Wiley, New York, 1960).
Cronin, J. W., *Bubble and Spark Chambers* (ed. Shutt, R. P.), Vol. I, p. 360 (Academic Press, New York, 1967).
Crouch, J. H., and Risk, W. S., Univ. Maryland Tech. Rep. 72-029, 1971 (AEC ORO-2504-176).
Crouch, J. H., and Risk, W. S., *Rev. Sci. Inst.* **43**, 632 (1972).
Crouch, J. H., and Risk, W. S., *Nuc. Inst. Meth.* **100**, 525 (1972).
Curran, S. C., Cockroft, A. L., and Angus, J., *Phil. Mag.* **40**, 929 (1949).
Curran, S. C., and Wilson, H. W., α, β, γ *Spectroscopy* (ed. Siegbahn, K.), Vol. 1, p. 303 (North Holland Publ. Co., Amsterdam, 1965).
Dahlgren, S., Kullander, S., and Lorenzi, R., *Nuc. Inst. Meth.* (submitted 1970).
Danby, G., Gaillard, J.-M., Goulianos, K., Lederman, L. M., Mistry, N., Schwartz, M., and Steinberger, J., *Phys. Rev. Lett.* **9**, 36 (1963).
Daniel, E. V., *Sov. Phys. Tech. Phys.* (*U.S.A.*) **10**, No. 4 (1965).
Davidenko, V. A., Dolgoshein, B. A., Semenov, V. K., and Somov, S. V., *Zhur. Eksp. i. Teoret. Fiz.* **55**, 426 (1968) [*Sov. Phys. J.E.T.P.* **28**, 223 (1969)].
Davidenko, V. A., Dolgoshein, B. A., and Somov, S. V., *Zhur. Eksp. i. Teoret. Fiz.* **55**, 435 (1968) [*Sov. Phys. J.E.T.P.* **28**, 227 (1969)].
Davidenko, V. A., Dolgoshein, B. A., Semenov, V. K., and Somov, S. V., *Coll. Int. Electronique Nucléaire* (Versailles, 1968).
Davidenko, V. A., Dolgoshein, B. A., Semenov, V. K., and Somov, S. V., *Nuc. Inst. Meth.* **67**, 325 (1969).
Davidenko, V. A., Dolgoshein, B. A., and Somov, S. V., *Nuc. Inst. Meth.* **75**, 277 (1969).
Davidenko, V. A., Dolgoshein, B. A., and Somov, S. V., *Sov. Phys. J.E.T.P.* **29**, 1 (1969).
Davidenko, V. A., Dolgoshein, B. A., Somov, S. V., and Starosel'tsev, V. N., *Sov. Phys. J.E.T.P.* **30**, 49 (1970).

Davidenko, V. A., Dolgoshein, B. A., Somov, S. V. and Starosel'tsev, V. N., *Sov. Phys. J.E.T.P.* **31**, 74 (1970).
Davidson, P. M., *Proc. Roy. Soc.* **A191**, 542 (1947).
Davier, M., Derado, I., Drickey, D., Fries, D., Mozley, R., Odian, A., Villa, F., and Yount, D., *Phys. Rev.* **D1**, 790 (1970).
Davis, H. T., Rice, S. A., and Meyer, L., *J. Chem. Phys.* **37**, 947 (1962).
Dean, A. J., Hutchinson, G. W., Ramsden, D., Taylor, B. G., and Wills, R. D., *Nuc. Inst. Meth.* **65**, 293 (1968).
Derenzo, S. E., Muller, R. A., Smits, R. G., and Alvarez, L. W., UCRL Report 19252 (1969).
Derenzo, S., Smadja, G., Smits, R., Zaklad, H., and Alvarez, L., *Nature*, **233** (1971).
Diambrini, G., and Giannini, I., *Nuc. Inst. Meth.* **41**, 56 (1966).
Dickey, F. R., *J. App. Phys.* **23**, 1336 (1952).
Dieperink, J., Kleinknecht, K., Steffen, P., Steinberger, J., Trippe, T., and Vannucci, F., CERN Internal Report N.P. 69–28 (1969).
Dimčovski, Z., private communication (1970).
Dimčovski, Z., Thesis, Univ. de Grenoble, 1970 and CERN Internal Report N.P. 70–30.
Dimčovski, Z., Favier, J., Charpak, G., and Amato, G., *Nuc. Inst. Meth.* **94**, 151 (1971).
Dimčovski, Z., CERN (ECFA) Report, p. 362 (1972).
Dolgoshein, B. A., Rodionov, B. U., and Luchkov, B. I., *Nuc. Inst. Meth.* **29**, 270 (1964).
Doviak, R. J., Lombardini, P. P., Moezzi, A., and Goldhirsh, J., *Nuc. Inst. Meth.* **48**, 344 (1967) and **54**, 161 (1967).
Druyvesteyn, M. J., and Penning, F. M., *Rev. Mod. Phys.* **12**, 87 (1940).
Eckardt, V., and Ladage, A., *Coll. Int. Electronique Nucléaire* (Versailles, 1968).
Eckardt, V., and Ladage, A., *Proc. Int. Conf. on Instrumentation for High-Energy Physics* (Dubna, 1970).
Eckardt, V., DESY Report, DESY 70/60 (1970).
Ehrmann, C. H., Fichtel, C. E., Kniffen, D. A., and Ross, R. W., *Nuc. Inst. Meth.* **56**, 109 (1967).
Eisenstein, B., Grossman, D., Sah, R., and Strauch, K., *Nuc. Inst. Meth.* **41**, 69 (1966).
El Bedewi, F., Goned, A., and Girgis, A. H., *J. Phys.* **A.5**, 292 (1972).
Emberson, D. L., Todkill, A., and Wilcock, W. L., *Electronics and Electron Physics*, **16**, 127 (1962).
Ermilova, V. K., Kotenko, L. P., Merzon, G. I., and Chechin, V. A., *Sov. Phys. J.E.T.P.* **29**, 861 (1969).

Eschstruth, P. T., Franklin, A. D., Hughes, E. B., Gibbs, B. W., Murphy, F. V., Reading, D. H., and Wright, K. E., *Nuc. Inst. Meth.* **63**, 96 (1968).
Evans, C. J., *Nuc. Inst. Meth.* **69**, 61 (1969).
Evans, L. R., and Grey, Morgan C., *Nature*, **219**, 713 (1968).
Evans, W. M., CERN Internal Report N.P. 68–27; and Private communication (1969).
Faissner, H., Ferrero, F., Ghani, A., Krienen, F., Novey, T. B., and Reinharz, M., *Nuc. Inst. Meth.* **20**, 161 (1963).
Faissner, H., Ferrero, F., Ghani, A., Krienen, F., Novey, T. B., and Reinharz, M., *Proc. Int. Conf. on Elementary Particles* (Sienna, 1963).
Falomkin, I. V., Kulyukim, M., Pontecorvo, D. B., and Sherbakov, Y. A., *Nuovo Cimento*, **34**, 1394 (1964).
Fakhry, A., *The Pyramids* (University of Chicago Press, Chicago, 1961).
Fano, U., *Phys. Rev.* **72**, 26 (1947).
Fazio, G. G., *Nature*, **225**, 905 (1970).
Fermi, E., *Phys. Rev.* **57**, 485 (1940).
Fichtel, C. E., Hartman, R. C., Kniffen, D., and Sommer, M., *Astrophys. J.* **171**, 31 (1972).
Fischer, J., and Zorn, G. T., *Rev. Sci. Inst.* **32**, 499 (1961).
Fischer, J., and Zorn G. T., *I.R.E.* NS–9, No. 3, 261 (1962).
Fischer, J., *I.E.E.E. Trans. Nuc. Sci.* NS–12, No. 4, 37 (1965).
Fischer, J., *Proc. Int. Conf. on Instrumentation for High-Energy Physics*, p. 31 (Stanford, 1966).
Fischer, J., *Proc. Int. Conf. on Electromagnetic Interactions* (Dubna, 1967).
Fischer, J., and Shibata, S., *Brookhaven Report*, BNL 14899 (1970).
Fitch, R. A., and Howell, V. T. S., *Proc. I.E.E.*, III, **4**, 849 (1964), and Patent No. 18136 (1961).
Fite, W. L., Rutherford, J. A., Snow, W. R., and van Lint, V., *Disc. Faraday Soc.* **33**, 264 (1962).
Focardi, S., Rubbia, C., Torelli, G., and Bella, F., *Nuovo Cimento*, **5**, 275 (1957).
Foeth, H., Hammarstrom, R., and Rubbia, C., *Nuc. Inst. Meth.* **109**, 521 (1973).
Fortune, R. D., Ph.D. thesis, University of Geneva (1960).
Fortune, R. D., CERN Reports, 1969 (ISR/DI/69–68) and (ISR/DI/69–73).
Fox, R. E., *J. Chem. Phys.* **26**, 1281 (1957).
Francis, G., *Ionization phenomena in gasses* (Butterworths Scientific Publications, 1960).

Frankel, S., Highland, V., Sloan, T., Van Dyck, O., and Wales, W., *Nuc. Inst. Meth.* **44**, 345 (1966).
Friend, B., *Nuc. Inst. Meth.* **65**, 311 (1968).
Frisch, O. R., Unpublished lectures (1948) (see West, 1953).
Frolich, A., *Nature*, **215**, 1362 (1967).
Fryberger, D., Horton, J., Jensen, D., Neumann, M., Nunamaker, T., Shea, T., and Telegdi, V., *I.E.E.E. Trans. Nuc. Sci.* **NS-15**, No. 3, 579 (1968).
Fukui, S., and Miyamoto, S., *Nuovo Cimento*, **11**, 113 (1959).
Fukui, S., and Miyamoto, S., *J. Phys. Soc., Japan*, **16**, 2574 (1961).
Fukui, S., Hayakawa, S., Tsukishima, T., and Nukushina, H., *Nuc. Inst. Meth.* **20**, 236 (1963).
Furry, W. H., *Phys. Rev.* **52**, 569 (1937).
Galaktionov, Y. V., Yech, F. A., and Lyubimov, V. A., *Nuc. Inst. Meth.* **33**, 353 (1965).
Galster, S., Gorres, J., Hartwig, G., Klein, H., Moritz, J., Schmidt-Parzefall, W., and Schopper, H., *Nuc. Inst. Meth.* **46**, 208 (1967).
Galster, S., Hartwig, G., Klein, H., Moritz, J., Schmidt, K., Schmidt-Parzefall, W., Schopper, H., and Wegener, D., *Nuc. Inst. Meth.* **76**, 337 (1969).
Garmire, G., and Kraushaar, W. L., *Space Science Reviews*, **4**, 123 (1965).
Garron, J. P., Grossman, D., and Strauch, K., *Rev. Sci. Instrum.* **36**, 264 (1965).
Garvey, J., Amato, G., Gildemeister, O., Briandet, P., Huffer, E., Appleby, L., Barnard, R., Lipman, N., Morris, D., Owen, D., White, D., Wilde, P., Lister, J., Proc. Int. Conf. on Instrumentation for High Energy Physics, p. 463 and p. 487 (Frascati, 1973).
Gelertner, H., *Nuovo Cimento*, **22**, 631 (1961).
Giacomich, R., and Lagonegro, M., *Nuc. Inst. Meth.* **34**, 347 (1965).
Gianelli, G., *Nuc. Inst. Meth.* **31**, 29 (1964).
Gluckstern, R. L., *Nuc. Inst. Meth.* **24**, 381 (1963).
Golutvin, I. A., Zanevsky, Y. V., Kiryushin, Y. T., Peshekhonov, V. D., Ryabtsov, V. D., and Sitnik, I. M., *Nuc. Inst. Meth.* **67**, 257 (1969).
Greinacher, H., *Helv. Phys. Acta.* **7**, 360 (1934).
Grey Morgan, C., 'Fundamentals of Electric Discharges in Gases' in *Handbook of Vacuum Physics* (Pergamon Press, Oxford, 1965).
Grey Morgan, C., CERN Report–NPA/Int. 67-7 (1967).
Grey Morgan, C., Private communication (1969).
Gromova, I., Nikanorov, V., Peter, G., and Pisarev, V., *Prib. Tekh. Eksp.* **1**, 64 (1965).
Grove, R., Kaufman, L., and Perez-Mendez, V., *Nuc. Inst. Meth.* **62**, 105 (1968).

Grove, R. L., Perez-Mendez, V., and Van Tuyl, R., *Nuc. Inst. Meth.* **70**, 306 (1969).
Grove, R., Ko, I., Leskovar, B., and Perez-Mendez, V., *Nuc. Inst. Meth.* **99**, 381 (1972).
Grunberg, C., Cohen, L., and Mathieu, L., *Nuc. Inst. Meth.* **78**, 102 (1970).
Gupta, S. L., and Saha, N. K., *Nuc. Inst. Meth.* **13**, 258 (1961).
Gygi, E., and Schneider, F., CERN Report, CERN 64-46 (1964).
Gygi, E., and Schneider, F., CERN Report, CERN 66-14 (1966).
Gygi, E., and Schneider, F., CERN Report-ISR/GS/69-46 (1969).
Gygi, E., and Schneider, F., CERN Report-ISR/GS/71-17 (1971).
Hampson, H. F., and Rastin, B. C., *Nuc. Inst. Meth.* **95**, 345 (1971) and **95**, 337 (1971).
Hampson, H. F., and Rastin, B. C., *Nuc. Inst. Meth.* **96**, 197 (1971).
Hanna, G. C., Kirkwood, D. H. W., and Pontecorvo, B., *Phys. Rev.* **75**, 985 (1949).
Harris, F., Katsura, T., Parker, S., Peterson, V. Z., Ellsworth, R., Yodh, G., Allison, W. W. M., Brooks, C. B., Mulvey, J. H., *Nuc. Inst. Meth.* **107**, 413 (1973).
Hasted, J. B., *Physics of Atomic Collisions* (Butterworths, London, 1964).
Healey, R. H., and Reed, I. W., *The Behaviour of Slow Electrons in Gases* (Amalgamated Wireless Ltd., Sydney, 1941).
Heitler, W., *The Quantum Theory of Radiation* (Oxford University Press, 1954).
Heintze, J., and Walenta, A. H., *Nuc. Inst. Meth.* (1973) (to be published).
Henning, P. G., Thesis (Hamburg, 1955).
Heyn, M. P. (1961), Unpublished. Quoted in Wenzel (1964).
Higinbotham, W. A., Jacobs, J. F., and Pate, H. R., *I.E.E.E. Trans. Nuc. Sci.* **NS-12**, No. 1, 386 (February 1965).
Hincks, E. P., Anderson, H. L., Evans, H. J., Fukui, S., Kessler, D., Klare, K. A., Lillberg, J. W., Sherbrook, M. V., Martin, R. L., and Kalmus, P. I. P., *Proc. Int. Conf. on Instrumentation for High-Energy Physics* (Stanford, 1966).
Hornbeck, J. A., *Phys. Rev.* **84**, 615 (1951).
Hornbeck, J. A., and Molnar, J. P., *Phys. Rev.* **84**, 621 (1951).
Horwitz, N. H., Lofstrom, J. E., and Forsaith, A. L., *Radiology*, **84**, 125 (1965) and *J. Nuc. Med.* **6**, 724 (1965).
Hough, J., and Drever, R. W., *Nuc. Inst. Meth.* **103**, 365 (1972).
Hough, J., *Nuc. Inst. Meth.* **105**, 323 (1972).
Hough, P. V. C., *Bubble and Spark Chambers* (ed. Shutt, R. P.), Vol. II, Chapter 3 (Academic Press, New York, 1967).

Howells, M. R., Osmon, P. E., and Sheldon, A. G., *Nuc. Inst. Meth.* **79**, 325 (1970).
Howells, M. R., and Osmon, P. E., *Jour. Phys. F.* **2**, 277 (1972).
Howells, M. R., and Osmon, P. E., *Jour. Phys. E.* **4**, 3 (1971).
Hübbeling, L., CERN Report, CERN 72-6 (1972).
Jesse, W. P., and Sadauskis, J., *Phys. Rev.* **97**, 1668 (1955).
Jones, B. D., Malos, J., Galbraith, W., and Manning, G., *Nuc. Inst. Meth.* **29**, 115 (1964).
Kanarek, T. I. *et al.*, *Proc. Int. Conf. High-Energy Accelerators and Instrumentation* (CERN, Geneva, 1959).
Kaufman, L., Perez-Mendez, V., Rindi, A., Sperinde, J. M., and Wollenberg, H. A., *Phys. Med. Biol.* **16**, No. 3, 417 (1971).
Kaufman, L., Perez-Mendex, V., Shames, D., and Stoker, G., *I.E.E.E. Trans. Nuc. Sci.* **NS-19**, No. 3, 169 (1972).
Kaul, W., Seyfried, P., and Taubert, R., *Z. Naturforsch.* **18A**, 431 (1963).
Keller, L. P., and Walschon, E. G., *Rev. Sci. Inst.* **37**, 1258 (1966).
Keller, L. P., Schluter, R. A., and White, T. O., *Nuc. Inst. Meth.* **41**, 309 (1966).
Keuffel, J. W., *Rev. Sci. Inst.* **20**, 202 (1949).
Kieffer, L. J., and Dunn, G. H., *Rev. Mod. Phys.* **38**, 1 (1966).
Kienzle, W., *Conference on Meson Spectroscopy* (Philadelphia, 1968).
Kirsten, F. A., Lee, K. L., and Conragan, J., *I.E.E.E. Trans. Nuc. Sci.* **NS-13**, No. 3, 583 (1966).
Klanner, R., *Diplomarbeit* (1969).
Klanner, R., Private communication (1971).
Klema, E. D., and Allen, J. S., *Phys. Rev.* **77**, 661 (1950).
Knasel, T. M., Walker, J. K., and Wong, M., *Rev. Sci. Inst.* **36**, 270 (1965).
Kobayashi, S., Itoh, H., and Yasumi, S., *Nuc. Inst. Meth.* **81**, 72 (1970).
Komarov, V. I., and Savchenko, O. V., *Nuc. Inst. Meth.* **34**, 289 (1965).
Korff, S. A., *Electron and Nuclear Counters* (Van Nostrand, New York, 1955).
Kotenko, L. P., Merzon, C. I., Vinogradov, A. D., and Vlasov, N. G., *Nuc. Inst. Meth.* **54**, 119 (1967).
Kozlov, M. I., and Rudenko, N. S., *Prob. Tekh Eksper.* **2**, 48 (1969) [*Instrum. Exper. Tech.* 319 (1969)].
Krienen, F., *Nuc. Inst. Meth.* **20**, 168 (1963).
Krienen, F., Private communication (1967).
Krienen, F., *Nuc. Inst. Meth.* **81**, 310 (1970).
Kuhlmann, W. R., Lauterjung, K. H., Schimmer, B., and Sistemich, K., *Nuc. Inst. Meth.* **40**, 118 (1966).

BIBLIOGRAPHY

Ladage, A., Private communication (1969).
Ladage, A., Dittmann, P., Eckardt, V., Horlitz, G., Joos, P., Kessler, G., Meyer, H., and Wolff, S., DESY/Unnumbered Report/1969.
Landau, L., *J. Phys., USSR*, **4**, No. 8, 201 (1944).
Lansiart, A. J., and Kellershohn, C., *Nucleonics*, **24**, No. 3, 56 (1966).
Lederman, L. M., *Rev. Sci. Inst.* **32**, 523 (1961).
Legler, W., *Brit. J. App. Phys.* **18**, 1275 (1967).
Lewis, T. A. D., and Wells, F. H., *Millimicrosecond Pulse Techniques* (Pergamon Press, Oxford, 1959).
Lichtenberg, G. C., *Novi. Comment. Gött*, **8**, 168 (1777).
Lillethun, A., Maglić, B., Stahlbrandt, C. A., Wetherell, A., Manning, G., Taylor, A. E., and Walker, T. G., CERN Report, CERN 64–30, p. 157 (1964).
Lillethun, E., and Zanella, P., CERN Report, CERN 64–30, p. 171 (1964).
Lindsay, J., and Pizer, I., *Coll. Int. Electronique Nucléaire* (Versailles, 1968).
Lipman, N. H. et al., *Dubna Conference* (1969), to be published.
Little, V. I., Mishra, S. R., and Rice-Evans, P., Unpublished (1966).
Loeb, L. B., *Basic Processes of Gaseous Electronics* (University of California Press, 1955).
Longo, M. J., Clark, J., Edict, G., Follebout, M., Gustafson, H., Keller, R., Sompayrac, L., and Young, K., *Nuc. Inst. Meth.* **95**, 53 (1971).
Lozanskii, E. D., *Sov. Phys.–Tech. Phys.* **13**, 1269 (1969).
Lozanskii, E. D., and Firsov, O. B., *Sov. Phys.–J.E.T.P.* **29**, 267 (1969).
Lozanskii, E. D., *Sov. Phys. Doklady*, **13**, 1134 (1969).
Lyubimov, V. A., and Pavlovsky, F. A., *Nuc. Inst. Meth.* **27**, 342 (1964).
Lyubimov, V. A., and Pavlovsky, F. A., *Nuc. Inst. Meth.* **27**, 346 (1964).
Maglić, B., *Proc. Int. Conf. on Instrumentation for High-Energy Physics*, p. 411 (Stanford, 1966).
Makowski, B., and Sadoulet, B., reported in Charpak et al. (1973).
Marinescu, L., Petrascu, M., and Voiculescu, G., *Nuc. Inst. Meth.* **54**, 327 (1967).
Manjavidze, Z. Sh., and Roinishvili, V. N., *Phys. Letters*, **24B**, 492 (1967).
Marshak, M. L., and Pruss, S. M., *Nuc. Inst. Meth.* **62**, 295 (1968).
Marshall, J., *Rev. Sci. Inst.* **25**, 232 (1954).
Martin, J. C., and Smith, I., Unpublished. Quoted in Bulos et al. (1967).
Marx, E., *Elektrotech. Z.* **45**, 652 (1924).

Massam, T., CERN Report, CERN 68-24 (1968).
McClure, G. W., *Phys. Rev.* **90**, 796 (1953).
McLaughlin, E. F., and Schafer, R. V., *Nuc. Inst. Meth.* **70**, 343 (1969).
Meek, J. M., and Craggs, J. D., *Electrical Breakdown in Gases* (Oxford University Press, 1953).
Menon, M., Naranon, S., Narasimham, V., Hinotani, K., Ito, N., Miyake, S., Craig, R., Creed, D., Osborne, J., and Wolfendale, A. W., *Proc. Roy. Soc.* **A301**, 137 (1967).
Merrill, F. H., and von Hippel, A., *J. App. Phys.* **10**, 873 (1939).
Meyer, D. I., and Terwilliger, K. M., *Rev. Sci. Inst.* **32**, 512 (1961).
Meyer, M. A., *Nuc. Inst. Meth.* **23**, 277 (1963).
Meyer, H., DESY Report, Fl–72–4.
Michael, D. N., and Schluter, R. A. (1963), Unpublished. Quoted in Cronin (1967).
Miller, D., and Hoffman, W., *Nuc. Inst. Meth.* **57**, 346 (1967).
Miller, L. S., Howe, S., and Spear, W. E., *Phys. Rev.* **166**, 871 (1968).
Mishra, S. R., Ph.D. thesis, University of London (1969).
Mistry, N., *Rev. Sci. Inst.* **32**, 526 (1961).
Miyamoto, S., *Nuc. Inst. Meth.* **30**, 361 (1964).
Moiseiwitsch, B. L., and Smith, S. J., *Rev. Mod. Phys.* **40**, 238
Morris Thomas, A., *Brit. J. Appl. Phys.* **2**, 98 (1951).
Möstl, K., and Timm, U., *Z. Phys.* **209**, 60 (1968).
Muller, R. A., Derenzo, S. E., Smadja, G., Smith, D. B., Smits, R. G., Zaklad, H., and Alvarez, L. W., *Phys. Rev. Letters*, **27**, 532 (1971).
Mulvey, J. H., International Conference, Frascati (1973).
Niel, M., Cassignol, M., Vedrenne, G., and Bonique, R., *Nuc. Inst. Meth.* **69**, 309 (1969).
Neumann, M. J., and Nunamaker, T. A., *Nuc. Inst. Meth.* **62**, 121 (1968) and *I.E.E.E. Trans. Nuc. Sci.* **NS–15**, No. 3, 591 (1968).
Neumann, M. J., and Nunamaker, T. A., *I.E.E.E. Trans. Nuc. Sci.* **NS–17**, No. 3, 43 (1970).
Odian, A., *Proc. Int. Conf. on Instrumentation for High-Energy Physics* (Stanford, 1966).
Ohasi, Y., *J. App. Phys., Japan*, **5**, 519 (1966).
Overseth, O. E. (1964), Unpublished. Quoted in Cronin (1967).
Pack, J. C., and Phelps, A. V., *Phys. Rev.* **121**, 798 (1961).
Pages, R., *Nuc. Inst. Meth.* **85**, 211 (1970).
Parker, J. H., and Lowke, J. J., *Phys. Rev.* **181**, 290 and 302 (1969).
Penning, F. M., *Electrical Discharges in Gases* (Philips, Eindhoven, 1957).
Perez-Mendez, V., and Pfab, J. M., *Nuc. Inst. Meth.* **33**, 141 (1965).
Peter, G., Tyapkin, A., Pisarev, A., and Tzou Chu-Lyang, *Nuc. Inst. Meth.* **20**, 201 (1963).

BIBLIOGRAPHY

Peyrou, C., *Bubble and Spark Chambers* (ed. Shutt, R. P.), (Academic Press, New York, 1967).
Phelps, A. V., Pack, J. L., and Frost, L. S., *Phys. Rev.* **117**, 470 (1960).
Pizer, I., CERN Report, CERN 64–30, p. 111 (1964).
Prasad, A. N., and Craggs, J. D., *Atomic and Molecular Processes* (ed. Bates, D. R.) (Academic Press, New York, 1962).
Prasad, Y., and Nath, N., *Nuc. Inst. Meth.* **77**, 254 (1970).
Przybylski, A., *Z. Phys.* **151**, 264 (1958).
Pullan, B. R., Howard, R., and Perry, B. J., *Nucleonics*, **24**, No. 7, 72 (1966).
Quercigh, E., *Nuc. Inst. Meth.* **41**, 355 (1966).
Raether, H., *Electron Avalanches and Breakdown in Gases* (Butterworths, London, 1964).
Ramana Murthy, P. V., and Demeester, G. D., *Nuc. Inst. Meth.* **56**, 93 (1967).
Rapp, D., and Briglia, D. D., *J. Chem. Phys.* **43**, 1480 (1965).
Rapp, D., and Englander-Golden, P., *J. Chem. Phys.* **43**, 1464 (1965).
Ratsey, O. L., *J. Inst. Elect. Engrs.* **93**, 3A, 245 (1946).
Reibel, K., and Schluter, R. A., *ANL Report*, KR/RAS—1. Unpublished (1963).
Reines, F., *Proc. Roy. Soc.* **A301**, 125 (1967).
Rice-Evans, P., and Mishra, S. R., Unpublished (1968).
Rice-Evans, P., *Jour. Phys. E.* **2**, 221 (1969).
Rice-Evans, P., and Mishra, S. R., *Nuc. Inst. Meth.* **67**, 337 (1969).
Rice-Evans, P., and Mishra, S. R., *Jour. Phys. E.* **4**, 638 (1971).
Rice-Evans, P., *Nature*, **232**, 625 (1971).
Rice-Evans, P., and Hassairi, I. A., *Phys. Lett.* **38A**, 196 (1972).
Rice-Evans, P., and Hassairi, I., *Nuc. Inst. Meth.* **106**, 345 (1973).
Rice-Evans, P., *Nuc. Inst. Meth.* **109**, 525 (1973).
Rice-Evans, P., to be published.
Rice-Evans, P., Hassairi, I., and Mishra, S. R., to be published (1973).
Riegler, A., *Brit. J. Appl. Phys.* **2**, 1423 (1969).
Rindi, A., Perez-Mendez, V., and Wallace, R., *Nuc. Inst. Meth.* **77**, 325 (1970).
Roberts, A., *Rev. Sci. Inst.* **32**, 482 (1961).
Roberts, A. (1963), Quoted by Astbury et al. (1967).
Roberts, A., *National Accelerator Laboratory 1969 Summer Study Report*, **SS–56**, Vol. III.
Rohrbach, F., Private communication (1971).
Roinishvili, V. N., Private communication (1969).
Rose, M. E., and Korff, S. A., *Phys. Rev.* **59**, 850 (1941).
Ross, R., Ehrmann, C., Fichtel, .C., Kniffen, D., and Ogelman, H., *I.E.E.E. Trans. Nuc. Sci.* **NS–16**, 127 (1969).

Rossi, B., *High-Energy Particles* (Prentice-Hall, New Jersey, 1965).
Rubbia, C., CERN/NP Report, 70–25/1970.
Rubbia, C., *N.P. Int. Rep.* 73-1, CERN (1973).
Rudenko, N. S., and Smetanin, V. I., *Sov. Phys. J.E.T.P.* **34**, 76 (1972).
Rutherford, E., and Geiger, H., *Proc. Roy. Soc.* **A81**, 141 (1908).
Rutherglen, J. G., and Paterson, J. M., *Rev. Sci. Inst.* **32**, 519 and 522 (1961).
Rutherglen, J. G., *Progr. Nuc. Phys.* **9**, 1 (1964).
Saro, S., and Srkalova, V., *Nuc. Inst. Meth.* **56**, 254 (1967).
Saudinos, J., Duchazeaubeneix, J. C., Laspalles, C., and Chaminade, R., *Nuc. Inst. Meth.* (1973) (to be published).
Schilly, P., Steffen, P., Steinberger, J., Trippe, T., Vannucci, F., Wahl, H., Kleinknecht, K., and Lüth, V., *Nuc. Inst. Meth.* **91**, 221 (1971).
Schlumbohm, H., *Z. Phys.* **151**, 563 (1958).
Schmied, H., Rohrbach, F., and Piuz, F., CERN Int. Report EMSA/TC–L/Int. 72–7 (see also CERN Courier, **11**, 281 (1971)).
Schmitt, F., Metzger, G., Gresser, J., Riedinger, M., and Sutter, G., *Nuc. Inst. Meth.* **76**, 258 (1969).
Schneider, F., and Höhne, K. H., *Nuc. Inst. Meth.* **20**, 152 (1963).
Schneider, F., CERN Report. AR/Int. GS/63–8 (1963).
Schneider, F., Private communication (1969).
Schnurmacher, G. L., *Nuc. Inst. Meth.* **36**, 269 (1965).
Schnurmacher, G. L., Clark, A. R., and Kerth, L. T., *Nuc. Inst. Meth.* **61**, 89 (1968).
Schübelin, P., Private communication (1968).
Schuller, E. G., Private communication (1969).
Schwab, A. J., *High-Voltage Measurement Techniques* (The M.I.T. Press, Massachusetts, 1972).
Seitz, F., *Phys. Fluids*, **1**, 2 (1958).
Seltzer, S. M., and Berger, M. J., *In studies in penetration of charged particles in matter*, NAS–NRC 1133, 187 (1964).
Singh, G., and Saha, N. K., *Nuc. Inst. Meth.* **33**, 9 (1965).
Springer, K., Poelz, G., and Daniel, H., *Nuc. Inst. Meth.* **69**, 240 (1969).
Steffen, P., and Vannucci, F., CERN N.P. Internal Report 69–29 (1969).
Stern, J. P., *Lichtenberg—A Doctrine of Scattered Occasions* (Thames and Hudson, London, 1963).
Sternheimer, R. M., *Phys. Rev.* **103**, 511 (1956).
Sternheimer, R. M., *Phys. Rev.* **115**, 137 (1959).
Sternheimer, R. M., *Phys. Rev.* **145**, 247 (1966).
Stetz, A. W., and Perez-Mendez, V., *Nuc. Inst. Meth.* **73**, 34 (1969).

BIBLIOGRAPHY

Stewart, A. T., and Roellig, L. O., *Positron Annihilation* (Academic Press, New York, 1967).
Sullivan, A. H., CERN Report, CERN 69-1 (1969).
Sutter, R. J., Bennett, G., Fischer, J., Friedes, J. L., Palevsky, H., Persson, R., Igo, G. J., and Simpson, W. D., *Nuc. Inst. Meth.* **54**, 71 (1967).
Swan, D. W., *Proc. Phys. Soc.* **78**, 423 (1961).
Swan, D. W., and Lewis, T. J., *Proc. Phys. Soc.* **78**, 448 (1961).
Symons, K. R., Harvard University thesis (1948).
Tan, B. C., Schmied, H., Rousset, A., Rohrbach, F., Piuz, F., Grey-Morgan, C., and Cathenoz, M., CERN Internal Report TC-L 70-7 (1970).
Tarlé, J., and Verweij, H., *Nuc. Inst. Meth.* **78**, 93 (1970).
Terhune, R. W., *Third Int. Symp. on Quantum Electronics*, Paris (1963).
Terwilliger, K. M., *Proc. Int. Conf. on Instrumentation for High-Energy Physics*, p. 389 (1966).
Tholl, H., *Z. Naturforsch.* **18a**, 587 (1963).
Thomson, J. J., and Rutherford, E., *Phil. Mag.* **42**, 392 (1896).
Townsend, J. S., *Proc. Roy. Soc.* **86**, 571 (1912).
Townsend, J., *Electrons in Gases* (Hutchinson, London, 1947).
Ullman, J. D., Bardin, R. K., Gollon, P. J., and Wu, C. S., *Nuc. Inst. Meth.* **66**, 1 (1968).
Uto, H., Yuan, L. C. L., Dell, G. F., and Wang, C. L., *Nuc. Inst. Meth.* **97**, 389 (1971).
Van Rossum, L., Private communication (1968).
Vasseur, J., Paul, J., Parlier, B., Leray, J. P., Forichon, M., Agrinier, B., Boella, G., Maraschi, L., Treves, A., Buccheri, R., and Scarsi, L., *Nature*, **226**, 534 (1970).
Vavilov, P. V., *Sov. Phys. J.E.T.P.* **5**, 749 (1957).
Verweij, H., International Conference, Frascati (May 1973).
Von Engel, A., *Ionized Gases* (London, Oxford University Press, 1965).
Wagner, K. H., *Z. Physik*, **189**, 466 (1966).
Walenta, A. H., Heintze, J., and Schürlein, B., *Nuc. Inst. Meth.* **92**, 373 (1971).
Walenta, A. H., *Nuc. Inst. Meth.* (1973, to be published) and *International Conference*, Frascati (1973).
Warren, R. W., and Parker, J. H., *Phys. Rev.* **128**, 2661 (1962).
Watten, J. W., Hopwood, W., and Craggs, J. D., *Proc. Phys. Soc.* **B63**, 180 (1950).
Welford, W. T., *Bubble and Spark Chambers* (ed. Shutt, R. P.), Vol. I, Chapter 5 (Academic Press, New York, 1967).
Wenzel, W. A., *Ann. Rev. Nuc. Sci.* **14**, 205 (1964).

Wenzel, W. A., *I.E.E.E. Trans. Nuc. Sci.* **13**, No. 3, 34 (1966).
West, D., *Progr. in Nuc. Phys.* **3**, 18 (1953).
West, D., *Proc. Phys. Soc.* **66A**, 306 (1953).
Weston, G. F., *Cold Cathode Glow Discharge Tubes* (Iliffe, London, 1968).
Wijsman, R. A., *Phys. Rev.* **75**, 833 (1949).
Wilkinson, D. H., *Ionization Chambers and Counters* (The University Press, Cambridge, 1950).
Wilkinson, K. J. R., *J. Inst. Elect. Engrs.* **93**, 3A, 258 (1946).
Wilkinson, K. J. R., *J. Inst. Elect. Engrs.* **93**, 3A, 1090 (1946).
Willis, B. A., and Grey-Morgan, C., *J. Phys.* **D2**, 1 (1969).
Willis, B. A., and Grey-Morgan, C., *Brit. J. Appl. Phys.* **2**, 1 (1969).
Willis, W. J., Majka, R., and Bergmann, W., *Nuc. Inst. Meth.* **91**, 29 (1971).
Willis, W. J., Hungerbuehler, V., Tanenbaum, W., and Winters, I. J., *Nuc. Inst. Meth.* **91**, 33 (1971).
Yasumi, Z., Itch, H., Kasaike, A., and Miyake, K., *Nuc. Inst. Meth.* **31**, 343 (1964).
Yuan, L. C. L., *I.E.E.E. Trans. Nuc. Sci.* **12**, No. 4, 206 (1965).
Yuan, L. C. L., Wang, C. L., and Prunster, S., *Phys. Rev. Lett.* **23**, 496 (1969).
Yuan, L. C. L., *Scintillation and Semiconductor Counter Symposium* (Washington D.C., March 1970).
Yuan, L. C. L., Uto, H., Dell, G. F., and Alley, P. W., *Phys. Lett.* **40B**, 689 (1972).
Zaklad, H., Thesis, Lawrence Radiation Lab., UCRL-20690 (1971).
Zanella, P., Lecture given at CERN, 1968.

Index

Accuracy (see Spatial resolution)
Alpha rays, 338
Alternating voltages, 306
Ambiguities, 100
Ambipolar diffusion, 52
Amplification, 252, 262
Analogue recording, 375
Angled tracks, 379
Archaeology, 357
A_2 resonance, 329
Attachment, 35, 190
Autotransformer, 184
Avalanche growth, 55, 246
Avalanche, space charge field, 60, 61
Avalanche statistics, 77

Balloon apparatus, 345
Bedford streamer chamber, 215
Blumlein line, 206, 221, 224
Boson spectrometer, 329
Breakdown, 74
Bremsstrahlung, 301, 335
Brightness of streamers, 224
Burial-chamber, 357

Camera, 129
Capacitance probe, 214
Capacitive memories, 163
Cardiac output, 357
Cascades of chambers, 284, 335
Cerenkov radiation, 332, 338
CERN avalanche chamber, 221
CERN drift chamber, 365
Characteristic impedance, 210
Charge exchange, 42
Charge transport, 80
Charpak chambers (see Proportional chambers)
Charpak prototype proportional chamber, 265
Chephren's Pyramid, 357
Chromatography, 353
Clearing field, 115, 117
Cloud chamber, 58

Clustering, 53
Coaxial cable, 91
Colliding protons, 330
Compton electrons, 343
Conducting plasma, 62, 68
Construction of
 drift chamber, 370
 proportional chamber, 265, 273
 spark chamber, 96, 103, 185
 streamer chamber, 215, 219
Cooling the electrons, 365
Corona discharge, 180
Cosmic rays, 344
Crab Nebula, 345
Critical length, 95, 116
Cross-section, 30
Current distribution—read-out, 160, 376
Cut-off method, 286
Cylindrical chamber, 97, 162
Cylindrical diffusion, 52
Cylindrical geometry, 252

Dead time in
 drift chamber, 380
 proportional chamber, 277
 spark chamber, 125
 streamer chamber, 224
Delay, 112, 202
Delay line, 357, 377
Delay line chamber, 168
Delta-rays, 25, 26, 280, 335
Density effect, 23, 289
DESY streamer chamber, 222
Dielectric, 211, 352
Diffusion, 47, 240
Diffusion coefficient, 51, 190, 239, 364
Diffusion in magnetic field, 54
Diffusion of electrons, 363
Digital recording, 135, 152, 185, 373
Diode memory, 165
Distinguishing particles, 243, 284
Double beta decay, 341
Double pulsing, 234

INDEX

Drift chambers, 360
Drift of electrons, 47
Drift velocity, 49, 59, 249, 361

$E \times B$ drift, 184, 281, 381
Effective temperature, 80
Efficiency for
  multiple tracks, 119
  single tracks, 116, 187, 275
Efficiency in
  drift chamber, 380
  proportional chamber, 275
  spark chamber, 116, 187
  streamer chamber, 232
Electrical breakdown, 347
Electric field, 245, 259
Electron affinity, 39
Electron average energy, 53
Electronic amplifiers, 293
Electron temperatures, 53, 365
Electrophoresis, 353
Electrostatic problems, 267
Elementary spark chamber, 89
Emission lines, 34
Energy loss, 19
Energy loss distribution, 29
Equivalent circuit, 106, 177
Erevan streamer studies, 238
Excitation, 29

Fermi surfaces, 342
Ferrite cores, 145
Fertile ancestor, 82
Fick's law, 50
Fiducial marks, 135
Film transport, 134
Finite size of spark, 138
Fluctuations in avalanches, 77, 254
Flying-spot digitizer, 129
Fractionally charged particles, 335
Freon, 99, 279

Gamma ray astronomy, 345
Gap length, 75
Gas discharge, 347
Gas—electronegative, 99, 272, 279
Gas—impurities, 98, 118, 220
Gas—optimum composition, 98, 123, 189, 220, 240, 269, 290, 362
Gas—purification, 100
Geiger counter, 43
Geiger regime, 246, 282
Geometric mean method, 286, 335

Goddard Space Flight Centre, 345
Green's theorem, 248

Harvard drift chamber, 370
Harvard wide-gap chamber, 184
Heidelberg drift chamber, 369
Helium chamber, 234
High pressure, 76, 127
High-voltage properties, 383
High-voltage pulse generation, 176, 182
High-voltage pulses, 203
Historical review, 82
Hough-Powell Device, 129
Hybrid chamber, 307
Hydrogen chamber, 236, 349
Hydrogen streamers, 236, 349
Hydrogen target, 220, 224

Identification of particles, 243, 284
Image intensifier, 229, 239
Inclined tracks, 187
Induced positive pulses, 298
Inductive location, 167
Intersecting Storage Rings (CERN), 330
Ion bombardment, 75
Ionization, 29
Ionization chamber, 249
Ionization
  effect on luminosity, 194
Ionization measurement, 237, 242, 284
Ionization potential, 30, 42
Insulator, 203
Isotropic response, 232

Kinematical analysis, 136
Kinetic theory, 47

Landau statistics, 26, 195, 237, 284, 335
Laplace transform, 178
Laser-excited chamber, 301
LC generator, 182
Least squares fitting, 194
Lecher's wires, 218
Left–right ambiguities, 379
Lenses, 130
Lichtenberg figures, 350
Likelihood ratio method, 286
Linearity, 254
Line broadening, 70
Liquid argon, 81, 318
Liquid xenon, 80, 319

INDEX

Low pressures, 282
Lozanskii theory, 68
Luminosity, 224

Magic gas, 270
Magnetic core read-out, 145
Magnetic field, 125, 148, 155, 184, 232, 281, 381
Magnetic tape, 170
Magnetostriction
  longitudinal mode, 149
  torsional mode, 155
Magnetostrictive read-out, 149, 185
Marx generator, 171, 176, 203, 207, 217, 221, 222, 224, 305
Maxwellian distribution, 45, 80
Mean free path, 45, 55
Medical diagnosis, 352
Memory time in
  drift chamber, 380
  proportional chamber, 277
  spark chamber, 122, 192
  streamer chamber, 202, 232
Metastable states, 41, 98
Michigan experiment, 344
Microphones, 137
Microwave chamber, 303
Minimum ionizing power, 21, 116, 291
Mirrors, 130
Mobility, 49
Momentum determination, 185
Moscow ionization studies, 238
Multiple scattering, 171, 232
Multiple sparks, 100, 155
Multiple tracks, 380

Narrow-gap chambers, 96
Negative ions, 35
Neon flash tube, 308
Neon/helium, 98
Neutrino, 328
Newton–Leibniz tangle, 83
Noble liquids, 80, 317
Noble molecular ions, 39
Noble solids, 80
Notre Dame, 89
Nuclear medicine, 356

On-line analysis, 100

Particle physics, 328
Penning effect, 41, 57, 98, 189, 234

Periodic wire chamber, 326
Photodiodes, 168
Photoelectric effect, 75
Photography, 96, 129, 134, 185, 227
Photoionization, 31, 67, 246, 310
Photon camera, 357
Photoproduction of hadrons, 331
Piezoelectric crystals, 139, 159
Plastic scintillator, 90, 335, 384
Plumbicon, 144, 166
Poisson distribution, 26, 237
Positron annihilation, 342
Precision (*see* Spatial resolution)
Prepulse in Blumlein, 211
Pressurised spark-gap, 181
Primary ionization, 23, 95, 203, 235, 238, 241, 242
Prisms, 130
Proportional chambers, 245
Proportional counter, 43, 245
Pulse-forming network, 205
Pulse measurement, 212
Pulse shaping, 203
Pulse (signal) formation in a proportional counter, 247, 262
Pulsing the chamber, 112

Quarks, 335
Quenching, 42, 70, 99, 270, 370

Radiative attachment, 37
Radioactive isotopes, 352
Raether criterion, 65, 93
Ramsauer–Townsend effect, 46, 99
RC differentiation, 252, 264
Read-out electronics, 293, 373
Recombination, 33, 61
Recovery time, 124, 192
Reflected binary code, 169
Reignited sparks, 194
Relativistic energies, 333
Relativistic rise, 20, 240, 242, 289, 291
Resistive probe, 213
Resonance capture, 37
Rise time, 204
Robbing, 113, 191
Rotating field, 305

Saclay drift chambers, 368
Secondary ionization coefficient, 75
Secondary processes, 74
Secondary sparks, 165

INDEX

Second coordinate, 376
Self-triggering, 338
Semiconductor detectors, 334
Sensitive time (*see* Memory time)
Series spark gap, 204, 218
Shunting spark gap, 178, 204, 218, 222
Simultaneous particles, 84
Simultaneous tracks, 190, 218
Skin depth, 214
SLAC streamer chamber, 219
Sonic detection, 137
Sound velocity, 137, 140
Space charge field, 222
Spark channel, 172
Spark counters, 322
Spark discharge, 83
Spark formation, 74, 92, 104
Spark formation time, 93
Spark gap, 91, 107
Sparkostriction, 158
Spark plasma, 124
Spatial resolution in
 drift chamber, 367, 370, 378
 proportional chamber, 280
 spark chamber, 103, 118, 136, 143, 148, 153, 194
 streamer chamber, 230
Spiral generator, 182
Spurious discharging, 96
Spurious pulsing, 84
Statistical fluctuations, 77, 334
Statistics of energy loss, 25
Stereo angle, 200
Stray capacitance, 204
Streamer brightness, 238
Streamer chambers, 196, 334
Streamer-cloud chamber, 311
Streamer growth, 60, 68, 196
Streamer photographs, 72
Streamer radius, 200
Streamer regimes
 end view, 198
 merged, 200
 side view, 198

Television camera, 140
Tension in wires, 269, 371
Termination, 205, 210
Third plane, 102
Threshold radius, 253
Thyratron, 106
Thyroid gland, 352
Time constant, 205
Time resolution (*see* Memory time)
Time stretching, 375
Total specific ionization, 23, 203
Townsend ionization coefficient, 55, 39
Townsend theory, 55
Track delineation, 171
Transition radiation, 332
Transmission line, 115, 155, 206, 222
Transparent electrodes, 198
Trigger amplifier, 91, 216
Triggered bubble chamber, 312
Triggering, 90, 176
Trigger system, 104

Ultra-violet photons, 349

Ventilatory function, 357
Vidicon camera, 141, 165
Villari effect, 149

Waveguide, 112, 304
Wide-gap spark chambers, 171
Wiedemann effect, 157
Wire instability, 267
Wire planes, 100, 145, 149
Work function, 43, 206

Xenon chamber, 357
X-rays, 333

Zero-crossing discriminator, 153

Spark, Streamer, Proportional

9780903840002.4